Refinery Byproducts as Raw Materials for Plasticizers

By

Sohini Mukherjee

ACKNOWLEDGEMENT

This book is an outcome of many years of effort where I have been accompanied and supported by many. The entire research work of the present book would have never been possible without the enormous support and immense guidance of **Dr. Mahua Ghosh, Associate Professor, Department of Chemical Technology, University of Calcutta.** I am extremely grateful to her for her encouragement in each and every step whenever I needed. I am extremely thankful to her for taking responsibility of my Ph.D. studies and for her generous support. I am obliged to her for her suggestions for improving the quality of my research.

I am extremely grateful to the **Head of the Department of Chemical Technology, University of Calcutta**, as well all the faculties and staff members of this department for their cooperation and support.

I would also like to express my gratitude to **Professor Dipak Kumar Bhattacharya, Ex-Ghosh Professor and Ex-Professor Emeritus, Department of Chemical Technology, University of Calcutta**, for his encouragement, constructive ideas and kind advice. His innovative ideas have enriched my study. It would not have been possible to do this work without his guidance and support.

I would like to thank **Prof. Aniruddha Mukhopadhyay** (Head of the Department), **Dr. Punarbasu Chaudhuri, Assistant Professor** and **Dr. Pritha Bhattacharjee, Assistant Professor**, as well as the non-teaching staff members of **Department of Environmental Science, University of Calcutta** for giving the opportunity to pursue my Ph.D under this Research Advisory Committee and providing me all the library facilities.

I would also like to express my gratitude to **Dr. Debasish Sarkar, Associate Professor, Department Chemical Engineering, University of Calcutta** for sharing his excellent knowledge in some part of my research work.

I am very much grateful to my all co-researchers of **Oil Laboratory, Department of Chemical Technology, University of Calcutta**, for their immense support. It has been a great experience to work with them and it would have been very difficult to complete my research work without their contributions.

I would also like to acknowledge **Centre for Research in Nanoscience and Nanotechnology (CRNN)** and **Department of Biotechnology-Interdisciplinary Life Science Programme (DBT-IPLS), University of Calcutta,** and **Bose Institute, S. N. Bose National Centre for Basic Sciences** for providing me with the necessary instruments for analyses.

I gratefully acknowledge the funding authorities "**Directorate of Vanaspati, Vegetable Oils and fats" (DVVOF), Ministry of Food and Consumer affairs** and **University of Calcutta** in pursuing my research work.

Warmhearted thanks to my beloved Parents **Mrs. Saptami Mukherjee** and **Mr. Sandip Kumar Mukherjee** for showering their blessings and love on me and always stand by me in each ups and downs during my research work. Heartfelt thanks to my husband **Mr. Subhajit Sil** for his enormous support and being there for me in each step. All of my family members deserve special thanks for being my side in the entire journey of the research and helping me in the successful completion of book.

Sohini Mukherjee

ABBREVIATIONS

CCAPL	Cetyl caprylate
CCAPR	Cetyl caprate
CMC	Carboxymethyl cellulose
HPMC	Hydroxylpropyl methyl cellulose
MC	methyl cellulose
KGM	konjac glucomannan
PEG	Polyethylene glycol
TEG	triethylene glycol
PG	propylene glycol
EA	Ethanolamine
GLY	Glycerol
RMIM	*Rhizomucor meihei* immobilized
TLIM	*Thermomyces lanuginosus* immobilized
NS435	*Candida antarctica* immobilized
CMC-GLY	Carboxymethyl cellulose + glycerol
CMC-CCAPL	Carboxymethyl cellulose + cetyl caprylate ester
CMC-CCAPR	Carboxymethyl cellulose + cetyl caprate ester
COFA	Castor oil fatty acid
FA	Furfuryl alcohol
FA-COFA	Ester of FA and COFA
DBP	Di-butyl phthalate
PVC+(DBP:FA-COFA)	Poly (vinyl chloride)+di-butyl phthalate+ ester of furfuryl alcohol and castor oil fatty acid
EC	Ethyl Cellulose

EC + DBP	Ethyl Cellulose+ di-butyl phthalate
EC + FA-COFA	Ethyl cellulose + ester of furfuryl alcohol and castor oil fatty acid
AFM	Atomic Force Microscopy
SEM	Scanning Electron Microscopy
TG	Thermo gravimetric analysis
DTA	Differential Thermal Analysis
DSC	Differential Scanning Calorimetry
XRD	X-ray diffraction
FTIR	Fourier-transform infrared spectroscopy
NMR	Nuclear magnetic resonance spectroscopy
TLC	Thin Layer Chromatography
GC	Gas Chromatography
WVP	Water vapor permeability
RSM	Response Surface Methodology
T	Temperature
W	Enzyme loading
M	Molar ratio
V	velocity of the reaction
V_m	Maximum reaction rate
K_m	Michaelis constant for the species FA and COFA
K_i	Inhibition coefficient of species FA and COFA
$K_{m(FA)}$	Michaelis constant for the species FA
$K_{m(COFA)}$	Michaelis constant for the species COFA
$K_{i(FA)}$	Inhibition coefficient of species FA
$K_{i(COFA)}$	Inhibition coefficient of species COFA

ABSTRACT

The book entitled "**Refinery Byproducts as Raw Materials for Plasticizers** "consists of two chapters. **Chapter 1** deals with the lipase catalyzed preparation of medium chain fatty acid based plasticizer for edible application. **Chapter 2** deals with the immobilized lipase catalyzed preparation of castor oil fatty acid based plasticizer for industrial applications and its kinetic study. Chapter 2 is again divided into two sections. **Section A** consists of the process parameter optimization of the esterification reaction along with characterization and performance evaluation of the ester as a biodegradable plasticizer. **Section B** deals with the kinetic study of the enzymatic esterification reaction of furfuryl alcohol and castor oil fatty acid.

Chapter 1: ***Enzymatic preparation of medium chain fatty acid (MCFA) based plasticizer for edible application***

The present study deals with the lipase catalyzed preparation of MCFA based plasticizers. Such as caprylic acid ($C_{8:0}$) and capric acid($C_{10:0}$) based esters of cetyl alcohol using immobilized *Candida antarctica*. The esters were characterized by flash point, slip melting point, viscosity and density analysis. Antimicrobial activities of the esters were also performed.

The esterification reactions were conducted with different molar ratios of caprylic acid or capric acid and cetyl alcohol (1:1,1:2 and 1:3) in a stirred tank batch reactor at different temperatures (45°C to 65°C) for different course of time with changing amounts of enzymes (4% to 12% w/w) of three different types of immobilized lipases (RMIM, NS 435 and TLIM). The substrate molar ratio, temperature and concentration of three different kinds of immobilized lipases (RMIM,NS 435 and TLIM) were variable experimental parameters. Cetyl caprylate (CCAPL) and cetyl caprate (CCAPR) esters were produced successfully under the optimized reaction conditions. The optimum reaction parameter for maximum yield of both CCAPL and CCAPR esters, were found to be observed at 6 h with the reactant molar ratio of 1:1 and 65^{0}C reaction temperature using 10%w/w immobilized NS 435.

Antimicrobial well diffusion study of the CCAPL and CCAPR esters was also done against the bacterial strain *E.coli* and *S.aureus.* CCAPL and CCAPR both exhibited better antimicrobial activity in case of *S.aureus* in comparison to *E.coli.*

Further aims of the study were to develop CMC-CCAPL and CMC-CCAPR ester composite films through emulsification and to determine the effect of cetyl caprylate (CCAPL) and cetyl caprate (CCAPR) ester on the moisture barrier, mechanical, thermal and crystalline and morphological properties of the glycerol-CCAPL/CCAPR plasticized CMC films.

XRD crystallography showed similar amorphous nature in case of both CCAPL and CCAPR esters. The esters in association with glycerol showed good thermal stability. The main physicochemical properties of the CMC films were affected by glycerol, CCAPL and CCAPR. The increased concentration of glycerol decreased the mechanical strength and increased the water vapor permeability. CCAPL and CCAPR could reduce the water vapor permeability of the CMC films. Therefore, an appropriate proportion of CCAPL or CCAPR and glycerol are required to reduce water vapor transmission through CMC films, while having no unfavorable effect on the mechanical properties of the films.

Chapter 2 - *Enzymatic preparation of castor oil fatty acid based plasticizer for industrial applications*

The study described the lipase catalyzed preparation of castor oil fatty acid based plasticizer for the application in polymeric material. The present chapter deals with the biocatalytic esterification of furfuryl alcohol (FA) and castor oil fatty acid (COFA) to produce a biodegradable, biorenewable and eco-friendly plasticizer. The present chapter is again divided in two sections i.e. Section A and Section B .

Section A:

This part of the study is concerned about the process parameter optimization of esterification reaction of furfuryl alcohol (FA) and castor oil fatty acid (COFA) as well as characterization and performance evaluation of the ester as a biodegradable plasticizer.

The ester was synthesized biocatalytically from furfuryl alcohol (FA) and castor oil fatty acid (COFA), using immobilized *Candida antarctica* Lipase B (NS 435) as bio catalyst. Reaction parameters were optimized and the ester (FA-COFA) obtained was characterized from their acid value, density, viscosity and flash point. The maximum yield of the ester was obtained at 3:1 molar ratio of FA and COFA at 60^0C reaction temperature with 10% w/w enzyme loading at 5 h. Formation of ester was confirmed by structure analysis using TLC, GC ,FTIR and ^{1}H-NMR , ^{13}C-NMRspectroscopy. The biodegradability of the ester was also confirmed by microbiological study.

The study also included the application of the ester as plasticizer into poly (vinyl chloride) (PVC) and ethyl cellulose film (EC) respectively.

The FA-COFA ester was used as secondary plasticizers to the PVC films. The PVC films were formulated using the combination of conventional plasticizer di-butyl phthalate and FA-COFA ester as secondary plasticizer at different concentration. Films were characterized by XRD, SEM, thermal analyses, mechanical performance and migration stability.PVC films with increasing FA-COFA (80% w/w) concentration showed amorphous nature of the film through XRD. SEM micrograph showed the compatibility of the FA-COFA ester with the PVC and DBP blend. Better mechanical strength was also observed with the increasing concentration of FA-COFA ester into the PVC film blend. Biodegradability study of the PVC films showed increased degradability with increasing concentration of the FA-COFA ester in the PVC film. The study showed that ester of FA and COFA could be a substitute of DBP as much as 80% (w/w) of the total plasticizer with improved elongation and tensile properties and such kind of sustainable resource based PVC blend films could be used as a good packaging material with biodegradable property.

Film characterizations of the FA-COFA ester plasticized ethyl cellulose (EC) films were evaluated by surface morphologies, XRD analysis, mechanical properties, thermal properties, water vapor permeability and migration stability test. It was an effective plasticizer with better mechanical properties and thermal stability at the increasing concentration of FA-COFA ester (15-25%w/w) containing EC film, than the traditional plasticizer, i.e. dibutyl phthalate (DBP) in producing good quality films. Chemical structure and the intermolecular interactions between FA-COFA ester and ethyl cellulose chains were the causative agents of these outstanding performances. Therefore, this FA-COFA ester, with significant plasticizing property, at a certain concentration, can be a substitute of DBP.

Hence the incorporation of the FA-COFA ester in PVC and EC film showed promising results, indicating potential commercial application.

Section B:

Kinetics of esterification reaction between furfuryl alcohol (FA) and castor oil fatty acid (COFA) using immobilized *Candida antartica* lipase B in a solvent free system was investigated in this study. Well known ping-pong bi bi mechanism was used to evaluate pertinent kinetic parameters. Experimental parameters (temperature, molar ratio of reactants and enzyme loading) were optimized to obtain maximum equilibrium yield of ester. Furthermore, functional variations of different kinetic parameters, present in the corresponding rate equation, such as the Michaelis-Menten constants, inhibition coefficients etc. were evaluated with respect to experimental parameters. The nature of the respective functional forms were quantitatively evaluated using the standard Response Surface Methodology (RSM).The conversion of COFA into ester was significantly influenced by operating parameters; the most suitable reaction conditions for maximization of products were obtained at 3.4:1 molar ratio FA and COFA, at 60°C reaction temperature with 10.21%w/w enzyme loading. The reaction mechanism/kinetics for esterification reaction was confirmed by fitting the experimental data with the proposed rate equation and a good fit with R^2value of 0.96.

GENERAL INTRODUCTION

Plasticizers are the low molecular weight non-volatile key components that are widely used in polymer industries as additives. The major role of such substance is to improve the flexibility and process ability of polymers by lowering the second order transition temperature and the glass transition temperature. According to the council of the IUPAC (International Union of Pure and Applied Chemistry), plasticizers act as ‘‘a substance or material incorporated in a material (usually a plastic or elastomer) to increase its flexibility, workability, or distensibility’’. The tension of deformation, hardness, density, viscosity and electrostatic charge are reduced by these substances. It also increases the polymer chain flexibility, resistance to breakage and dielectric constant. Other affected physical properties include degree of crystallinity, electric conductivity, flash point and resistance to biological degradation.

During the last decade, the global manufacture of plasticizers was around 5 million tons per year. Plasticizers were applied to about 60 polymers and more than 30 groups of products. The use of plasticizers for manufacture of plastic products is not a new practice. Its application to alter the polymer characteristics was initiated in the 1800s. Previously, there was a trend of using natural camphor oil based celluloid or celluloid lacquers for plasticization purposes, but that was unsatisfactory for many end uses. Later, triphenyl phosphate was experimented to substitute camphor oil, representing the beginning of the ester plasticizers era. Initially, phthalic acid esters found applications as plasticizers and continue to be the leading class of plasticizers in the 21st century.Since 1930s, the most widely used plasticizer was di(2-ethylhexyl) phthalate (DEHP) which was known as dioctyl phthalate (DOP). Plastic products with huge variety and its numerous applications have headed to the development of upgraded plasticizers in order to meet product quality and

OBJECTIVES

The present study aims at investigating:

1. The enzymatic preparation of medium chain fatty acid based plasticizer for edible applications.
2. The enzymatic preparation of castor oil fatty acid based plasticizer for industrial applications.
3. The enzyme kinetics for esterification between furfuryl alcohol and castor oil fatty acid .

CONTENTS

Section B:

Kinetics of esterification reaction between furfuryl alcohol (FA) and castor oil fatty acid (COFA) using immobilized *Candida antartica* lipase B in a solvent free system was investigated in this study. Well known ping-pong bi bi mechanism was used to evaluate pertinent kinetic parameters. Experimental parameters (temperature, molar ratio of reactants and enzyme loading) were optimized to obtain maximum equilibrium yield of ester. Furthermore, functional variations of different kinetic parameters, present in the corresponding rate equation, such as the Michaelis-Menten constants, inhibition coefficients etc. were evaluated with respect to experimental parameters. The nature of the respective functional forms were quantitatively evaluated using the standard Response Surface Methodology (RSM).The conversion of COFA into ester was significantly influenced by operating parameters; the most suitable reaction conditions for maximization of products were obtained at 3.4:1 molar ratio FA and COFA, at 60°C reaction temperature with 10.21%w/w enzyme loading. The reaction mechanism/kinetics for esterification reaction was confirmed by fitting the experimental data with the proposed rate equation and a good fit with R^2value of 0.96.

GENERAL INTRODUCTION

Plasticizers are the low molecular weight non-volatile key components that are widely used in polymer industries as additives. The major role of such substance is to improve the flexibility and process ability of polymers by lowering the second order transition temperature and the glass transition temperature. According to the council of the IUPAC (International Union of Pure and Applied Chemistry), plasticizers act as ''a substance or material incorporated in a material (usually a plastic or elastomer) to increase its flexibility, workability, or distensibility''. The tension of deformation, hardness, density, viscosity and electrostatic charge are reduced by these substances. It also increases the polymer chain flexibility, resistance to breakage and dielectric constant. Other affected physical properties include degree of crystallinity, electric conductivity, flash point and resistance to biological degradation.

During the last decade, the global manufacture of plasticizers was around 5 million tons per year. Plasticizers were applied to about 60 polymers and more than 30 groups of products. The use of plasticizers for manufacture of plastic products is not a new practice. Its application to alter the polymer characteristics was initiated in the 1800s. Previously, there was a trend of using natural camphor oil based celluloid or celluloid lacquers for plasticization purposes, but that was unsatisfactory for many end uses. Later, triphenyl phosphate was experimented to substitute camphor oil, representing the beginning of the ester plasticizers era. Initially, phthalic acid esters found applications as plasticizers and continue to be the leading class of plasticizers in the 21st century.Since 1930s, the most widely used plasticizer was di(2-ethylhexyl) phthalate (DEHP) which was known as dioctyl phthalate (DOP). Plastic products with huge variety and its numerous applications have headed to the development of upgraded plasticizers in order to meet product quality and specification requirements. By the past half century, legislation regarding the health and safety issues has led to the development of an extensive range of currently available commercial plasticizers. Fatty acid esters, such as tartrates, benzoates, and esters of adipic, azelaic and sebacic acid, chlorinated hydrocarbons are included. The demand for plasticizers increased with the continuous growth of plastic industry. There are innumerable choices of plasticizers which are now offered by the current market that can be selected for specific applications to meet critical material requirements. There are growing concerns and controversy, regarding the use of phthalates and their toxic effects on the human health and the environment from the past 1980s. Thus, the use of plasticizers is being questioned due to their possible toxicity problems, related to the migration of phthalates. This matter led to the development of new and restrictive regulations regarding the use of phthalates in flexible PVC products. Currently, there is a trend towards replacing DOP by either di-isononyl phthalate (DINP) or di-isodecyl phthalate (DIDP), which are higher molecular weight phthalates and therefore are more permanent, have lower solubility and present slower migration rates. Additionally, other unconventional plasticizers, with lower toxicity and lesser migration levels have been extensively used in the last eras to overcome these difficulties and these kinds of plasticizers might be very suitable for applications which are specially sensitive to this phenomenon.

Currently, the increasing interests regarding the use of renewable raw material based plasticizers are characterized by low toxicity and low migration. This particular kind of plasticizer includes epoxidized triglyceride vegetable oils from linseed oil, soybean oil, sunflower oil, castor-oil, and fatty acid esters (FAEs). This hunt for renewable resource based plasticizers is also associated to the increased interest of material researchers and industries in the development of new bio-based materials, made from biodegradable resources with the potential to reduce the use of conventional plastic goods. It is very much realistic to suppose that plasticizers for biopolymers should preferably also be biodegradable.

In this respect, most of the traditional plasticizers used in synthetic polymer processing are not suitable for some biodegradable thermoplastics such as poly(3-hydroxybutyrate) (PHB), reinforcing the need for further investigations and developments in this area. Though a total replacement of conventional plasticizers by the renewable one is just impossible, at least for some specific applications such a replacement seems obvious and useful. Hence the book aims at synthesizing nontoxic and biodegradable plasticizers.

Furthermore, modern enzymology has achieved improvements in the development and application of lipase as catalyst. New immobilization techniques make possible to use enzymes in industrial processes in a similar way to the classical catalysis for heterogeneous reactions. However, the synthesis is classified as chemical or enzymatic production according to the catalysts employed in the process. Higher yields can be obtained in lesser time when chemical transesterification is applied. Though, there are disadvantages such as high energy requirements, difficulties in the recovery of the catalyst and potential pollution to the environment are major disadvantages in alkali or acid catalyzed processes. Lipases can catalyze a reaction in mild conditions and an increasing number of researchers have reported on its use. One of the blockages for industrial application of lipase is the high cost. Introduction of immobilization techniques can improve the lipase stability and allow for repeated utilization. Lipases are widely employed to catalyze hydrolysis, alcoholysis, esterification and transesterification of carboxylic esters. Lipases have excellent catalytic activity and stability in non-aqueous media, which facilitate the esterification and transesterification process during plasticizer production. So this study is adapting enzymatic esterification method for the production of bioplasticizer.

OBJECTIVES

The present study aims at investigating:

1. The enzymatic preparation of medium chain fatty acid based plasticizer for edible applications.
2. The enzymatic preparation of castor oil fatty acid based plasticizer for industrial applications.
3. The enzyme kinetics for esterification between furfuryl alcohol and castor oil fatty acid .

CONTENTS

ABSTRACT

STUDIES ON ENZYMATIC PREPARATION OF PLASTICIZERS FOR EDIBLE AND INDUSTRIAL APPLICATION FROM VEGETABLE OILS AND REFINERY BYPRODUCTS

Plasticizers have wide applications in edible as well as non-edible sectors. Lipid derived plasticizers are not only based on renewable sources, its biodegradability index may also quite appreciable. Chapter 1 of the present book describes the enzymatic esterification of caprylic and capric acid with cetyl alcohol respectively to produce edible grade plasticizer and the Chapter 2, section A includes the enzymatic esterification of furfuryl alcohol (FA) and castor oil fatty acid (COFA) for the industrial grade plasticizer. Kinetics of esterification of FA and COFA was also deduced and explained by Michaelis-Menten equation in Chapter 2, section B. The detailed physicochemical characterizations of the cetyl caprylate (CCAPL), cetyl caprate (CCAPR) esters and FA-COFA esters were investigated. Further study confirmed the antimicrobial activity and biodegradability of edible and non-edible plasticizers respectively. Afterwards the evaluations of esters as edible grade plasticizers were performed by applying into the carboxymethyl cellulose film (CMC) blend in association with glycerol through emulsification. FA-COFA ester was also applied as plasticizer into two kinds of polymeric blend separately i.e.Poly(vinyl chloride) (PVC) and ethyl cellulose (EC) by solvent cast method and in both the cases of the polymers, the performances of the FA-COFA ester were compared with the conventional plasticizer i.e. di-butyl phthalate (DBP). Each film irrespective of being edible or inedible was characterized by XRD, SEM, thermal analysis, mechanical performance. Both the CMC-CCAPL and CMC-CCAPR composite films showed reduced water vapor transmission and good thermal stability whereas the increased concentration of glycerol decreased the mechanical strength and increased the water vapor permeability.Therefore, an appropriate proportion of glycerol and CCAPL/CCAPR are required to build a plasticized edible film.

PVC and EC films containing FA-COFA ester in higher concentration showed impressive polymer-plasticizer compatibility through SEM and XRD. Improved mechanical strength and migration stability of both PVC and EC films were observed with the increasing concentration of the FA-COFA ester in the film blends. PVC film degradation was also confirmed by the *in vitro* degradation study. Therefore, this FA-COFA ester, with significant plasticizing property, at a certain concentration, can be a substitute of petroleum based plasticizer.

Kinetic data indicates that the esterification of FA with COFA catalyzed by *Candida antarctica* lipase B can be described adequately by a Ping Pong bi–bi mechanism with competitive inhibition.

CHAPTER 1

ENZYMATIC PREPARATION OF MEDIUM CHAIN FATTY ACID BASED PLASTICIZER FOR EDIBLE APPLICATION

INTRODUCTION

Increased awareness of environmental concerns is some of the causes that have encouraged the growing interest in developing edible films from biodegradable materials. There are several types of biopolymers, including proteins, polysaccharides and lipids or their combinations, which have been extensively studied to prepare edible coatings or films.

As a renewable resource, cellulose is very much abundant and its derivatives have excellent film-forming properties. Edible films based on cellulose derivative with low water content are very efficient barriers to oxygen and aroma compounds. Water soluble cellulose derivatives such as methylcellulose (MC), hydroxylpropyl methyl cellulose (HPMC) and carboxymethyl cellulose (CMC) are of interest to researchers since they are able to form a continuous matrix. These types of ethers of cellulose have no adverse effects on human health, and are highly efficient additive to improve the product and processing properties in various fields of application, from cosmetics, food products, and pharmaceuticals to products for the paper and textile industries. Regardless of the great number of studies about the film-forming properties of MC and HPMC edible films, only few works about the film-forming properties of CMC have been reported. The effect of murta (*Ugni molinae Turcz*) leaves extract on some physical properties of CMC-based films was discussed in one study.

CMC exhibits excellent film forming properties, thermal gelation and efficient oxygen, carbon dioxide, and lipid barrier properties. Henceforth, it is extensively used in the pharmaceutical and food industries. However, water vapor transmission of CMC is very poor. Nonetheless, its water vapor barrier properties might be upgraded by adding hydrophobic materials, such as lipids, into the film forming solutions. Films prepared by lipids consist of a primary function of blocking moisture transport due to their high hydrophobicity and reducing abrasion of the food surface during handling and transport. Fatty acids and its derivatives can also act as plasticizers into different biopolymer films. Effect of different fatty acids on maize starch and gelatin film has already been reported. Lipid addition decreases water vapor permeability rate of biopolymer films which is dependent upon the hydrophobic/hydrophilic ratio provided by their components. Incorporation of lipid components into hydrophilic films may have hostile effects on the mechanical and optical properties of edible films. The effects of palm olein on the physical properties of the konjac glucomannan (KGM)–CMC blend films were studied and the study showed that CMC with deacetylated KGM was able to increase emulsion stability during drying and emulsion films were produced with better barrier and mechanical efficiency.

Normally plasticizers are added to polymers to reduce brittleness, since they increase the free volume between polymer chains, decreasing intermolecular forces and thus increasing flexibility and extensibility of polymers. Many researchers studied the effects of various polyols on starch-based films. The most preferred polyols were glycerol, sorbitol, and PEG. However, plasticization, due to the involvement of water and plasticizers, such as polyols or lipids, induces an increase of gas and vapor transfer through such edible films. Glycerol is widely used as plasticizers in the preparation of edible films. The fatty acid esters of cetyl alcohol are known as wax esters which contain a good barrier property and have wide usage in the cosmetic industry. They are used as additives in lubricants, polishes, plasticizers and coating materials. But there has been no report on the application of fatty acid ester of cetyl alcohol as plasticizer into the CMC films. Henceforth, an appropriate quantity of glycerol and fatty acid ester derivative is required in the film solution to reduce water vapor transmission through the resulting edible films. At the same time, their occurrence should, preferably, not have any negative effect on the mechanical and optical properties of the films.

The present study deals with the lipase catalyzed preparation of the cetyl alcohol esters of caprylic acid ($C_{8:0}$) and capric acid ($C_{10:0}$) by *Candida antarctica* with optimized process parameters. The esters were characterized by flash point, slip melting point, viscosity and density analysis. Antimicrobial activities of the esters were also performed. Further aims were to develop CMC/lipid composite films through emulsification and to determine the effect of cetyl caprylate (CCAPL) and cetyl caprate (CCAPR) ester on the moisture barrier, mechanical, thermal and crystalline and morphological properties of the glycerol-CCAPL/CCAPR plasticized CMC films.

OBJECTIVES

The present chapter includes:

1. The lipase catalyzed syntheses of ester of medium chain fatty acids ($C_{8:0}$& $C_{10:0}$) and cetyl alcohol with immobilized lipases from *Rhizomucor meihei* (RMIM), *Candida antarctica* (NS 435), and *Thermomyces lanuginosus* (TLIM).

2. After the preparation of the esters, following analyses were performed-

   (a) Characterization of ester by FTIR, viscosity and density etc.

   (b) Antimicrobial activity of the esters

3. Application of esters as plasticizer in carboxymethyl cellulose (CMC) films and their performance evaluation.

REVIEW OF LITERATURE

Recently, the trend of the environmental research is moving towards the replacement of petroleum based plastics and plasticizers, in an economical manner, with biodegradable materials which offer a competitive mechanical properties. Biopolymers are regarded as the most potent materials for this purpose [1]. Though, they are usually poor in mechanical performance due to their fragility and brittleness which can limit their potential for application. Henceforth, plasticizers are introduced as additive to provide the necessary workability to biopolymers. This kind of products gained much attention when biodegradable additives and plasticizers also became the focus of material scientists [2]. The practice of ecofriendly or biodegradable plasticizers, with less toxicity and good compatibility with several plastics, resins, rubber and elastomers in substitution of conventional plasticizers, such as phthalates and other kinds of synthetic plasticizers which attracted the market along with the increasing worldwide trend towards use of biopolymers. Application of renewable resource based polymeric films depends on several characteristics including cost availability, functional attributes, mechanical properties, gloss and opacity, barrier properties, structure resistance to water and sensorial acceptance [3]. These characteristics are greatly influenced by parameters such as the type of material used as structural matrix (conformation, molecular mass, charge distribution), film manufacturing conditions (solvent, pH, concentration, temperature, etc.) and the type and concentration of additives (plasticizers, crosslinking agents, antimicrobials, antioxidants, etc.).The bio-polymeric films often get dehydrated and produce sturdy cohesive films that usually need plasticizers. Plasticizer reduces intermolecular forces amongst the polymer chains, which helps to improve the flexibility and chain mobility.Current research have focused on the use of polyols such as glycerol [4-10], ethylene glycol, diethylene glycol, triethylene glycol

(TEG), tetraethylene glycol and polyethylene glycol (PEG) ,propylene glycol (PG) [11-19], sorbitol [20-32], mannitol[33] and xylitol [30,34]; fatty acids [35-40]; monosaccharides (glucose, mannose, fructose, sucrose) [41-44]; ethanolamine (EA) [45]; urea [46];triethanolamine (TEA) [47]; vegetable oils; lecithin; waxes [48,49]; amino acids [50]; surfactants [51] and water [52] as plasticizers of edible and biodegradable films. There are also disadvantages of addition of plasticizers in respect to the edible films. In most of the cases the issues are regarding the increased level of gas, solute and water vapor permeability and the decrease in cohesion affects mainly mechanical properties. Hence, the biopolymers film properties are based on equilibrium between the extent of cross linking of the polymer matrix and the application of exact amount of plasticizers for better workability [53-55]. The plasticizer can surpass the compatibility limit with the biopolymer beyond a critical concentration level. Due to the hygroscopic nature of the biopolymers and bio-plasticizers, film moisture content is exaggerated by ambient conditions [56].

Generally, the edible and biodegradable films are the base materials used to form by polysaccharides, proteins and lipids compounds. Polysaccharides have good film-forming properties, providing efficient barriers against oils and lipids, although their moisture barriers are poor [57]. Protein-based films have highly interesting properties: mechanical and barriers properties are generally better than those of polysaccharides based films [58]. Lipid compounds have been used to make edible films for their excellent moisture barrier properties, but they can cause textural and organoleptical problems. In the last years, several studies have been dedicated to form composite films, by combining one or several lipid compounds with one hydrocolloid-based structural matrix [59-61].Commonly, polysaccharide films are obtained from starch, alginate, cellulose ethers, chitosan, carrageenan, or pectins and show good gas barrier properties. Linear structure of some of these polysaccharides, for example, cellulose (1,4- b-d-glucan), amylose (a component of starch, 1,4-a-d-glucan), chitosan (1,4-b-d-glucosamine polymer), renders their films tough, flexible and transparent. Their films are resistant to fats and oils [62]. Most available renewable resource is cellulose and its derivatives which have excellent film-forming properties. Cellulose derivative based edible films containing low water content are very efficient barriers to oxygen and aroma compounds [63,64].Methyl cellulose (MC), hydroxy propyl methyl cellulose (HPMC) and carboxymethyl cellulose (CMC) are the water soluble derivatives of cellulose which are of great interest to researchers since they are able to form a continuous polymeric matrix. CMC is an ether of cellulose which shows thermal gelation, excellent film forming ability. For this reason, it is widely used in the pharmaceutical and food industries. There are a number of studies about the film-forming properties of HPMC [65-67] and MC [68-70] edible films, very few works about the film-forming properties of CMC has been reported. Bifani et al. [71] described the effect of murta (*Ugni molinae Turcz*) leaves extract on the physical properties of CMC-based films. In addition, the physical properties of blend films produced from CMC–starch [72], pullulan–alginate [73] and konjac glucomannan (KGM) [74] have been investigated. Chai and Isa[75] studied the effect of oleic acid on the carboxymethyl cellulose based biopolymer electrolyte. Almasi described the physicochemical properties of starch-CMC-nanoclay biodegradable films[76].Chai and Isa also invented a proton conducting solid-bio polymer electrolyte which is based on oleic acid and glycerol doped CMC[75].Tongdeesoontorn studied the effect of CMC concentration on physical properties of biodegradable cassava starch based films[77]. Ayatunga Arikkibar and Ferhundeüs studied thermal, mechanical and water adsorption properties of corn-starch –CMC-MC biodegradable films [78].Tong et.al. described the preparation and characterization of pullulan–alginate–CMC blend films[73].

The use of plasticizers is necessary to improve the mechanical properties of protein or polysaccharide-based edible films. However, plasticization, due to the involvement of water and plasticizers, such as polyols or lipids, induces an increase of gas and vapor transfer through such edible films [79]. Glycerol is one of the commonly used edible plasticizers in the preparation of edible films. Accordingly, an appropriate proportion of plasticizer and fatty acid is required in the film solution to reduce water vapor transmission through the resulting edible films and their existence should, ideally, not have any adverse effect on the mechanical properties of the films. There are very least amount of work has been done on the effect of lipid on the properties of biopolymer film. Ghanbarzadeh described the effect of oleic acid on the physical and mechanical properties of CMC based films[1].

Medium-chain fatty acids (MCFA) are colorless, transparent, tasteless, and low-viscous 'water-like' liquid at normal temperature and include 8-10 carbon fatty acids[80]. MCFAs are markedly different from long-chain fatty acids with regard to physical properties, digestion/absorption, biodegradation, and body fat accumulation. Its utilization as an edible oil started in the 1970s, and it is currently used as an additive for foods, base material of pigments, and mold and lubricating oil in food production. Present study includes caprylic acid and capric acid in the production of edible grade plasticizer which can be added in the food coatings. Fakhouri described the effects of fatty acids over the physical properties of gelatin and gluten based edible films [81].The concept is to synthesize ester of caprylic acid and capric acid biocatalytically in combination with cetyl alcohol which is known as wax esters. Generally, they have carbon chain lengths of 6-10 or more. Because of their good emulsifying, stabilizing, and detergent-like properties, wax esters are frequently used in industry as carriers of water-insoluble materials for application onto various surfaces. Wax esters are biodegradable, biocompatible, nontoxic, and have excellent wetting capabilities. These kinds of wax esters are known to have a wide range of applications as lubricants, polishes, plasticizers, coating materials in the pharmaceutical and food industries [80].These kinds of ester are generally produced by chemical or enzymatic esterification reaction. Recent studies showed biocatalytic production of wax esters. Lipase catalyzed synthesis of wax ester was obtained by the esterification of palm stearine and cetyl alcohol [82].The reaction was carried out at 30°C reaction temperature, 300IU/ml enzyme concentration and 0.21g/ml substrate concentration. Kuo described the lipase catalyzed production of cetyl octanoate in super critical carbon di oxide[83].

MATERIALS AND METHODS

LIPASE CATALYZED ESTERIFICATION REACTION OF CAPRYLIC ACID WITH CETYL ALCOHOL

(i)Materials for the production of cetyl caprylate (CCAPL) ester

Caprylic acid ($C_{8:0}$) was procured from Merck India Ltd., and its composition was determined by gas chromatography (GC). Cetyl alcohol was purchased from LobaChemie.,India.

Three types of enzymes e.g. *Rhizomucor meihei* (RMIM),*Candida antarctica* (NS 435),*Thermomyces lanuginosus* (TLIM) were used in this study. These lipases were immobilized using macroporous acrylic resin with a water content of 10 % (w/w), and obtained from Novozyme India Pvt. Ltd, Bangalore, India.

All other chemicals and solvents were of analytical grade and obtained from Merck, India Ltd.

(ii)Method

The reaction was performed with different molar ratios of caprylic acid and cetyl alcohol (1:1,1:2 and 1:3) in a flat bottom flask and stirred at 200 rpm at different temperatures (45^{0}C to 65^{0}C) for different course of time with changing amounts of enzymes (4% to 12% w/w) of three different types of immobilized lipases (RMIM, NS 435 and TLIM).The reaction was performed in a complete solvent free environment and under vacuum (Pressure: 28mm Hg). The substrate molar ratio, temperature and enzyme loading of three different kinds of immobilized lipases (RMIM,NS 435 and TLIM) were variable experimental parameters and optimized with a single reaction mixture that were studied one at a time and eventually followed in the subsequent reactions. The effect of three different lipases on the yield of the ester was also studied in this study. The entire reaction procedure has been provided in the schematic diagram **[Scheme 1].**

The reaction course was monitored by determining intermittent acid value of the product with time. The ester formation was confirmed by thin layer chromatography (TLC), and the percent yield of ester and fatty acid composition of the ester was determined simultaneously.

After alkali refining and steam distillation process the ester content and the absence of free fatty acid and alcohol was confirmed by thin layer chromatography.

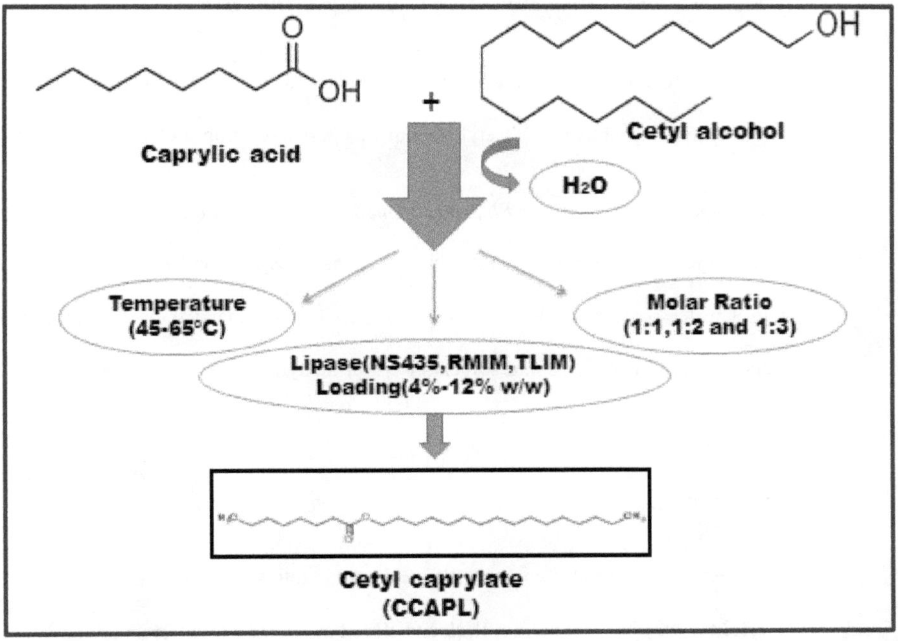

Scheme 1.Reaction mechanism for the production of CCAPL from caprylic acid and cetyl alcohol

LIPASE CATALYZED ESTERIFICATION REACTION OF CAPRIC ACID WITH CETYL ALCOHOL

(i) Materials for the production of cetyl caprate (CCAPR) ester

Capric acid ($C_{10:0}$) was procured from Merck India Ltd., and its composition was determined by gas chromatography (GC). Cetyl alcohol was purchased from Loba Chemie., India.

Three types of enzymes e.g. *Rhizomucor meihei (*RMIM) , *Candida antarctica* (NS 435), *Thermomyces lanuginosus (*TLIM) that are immobilized using macroporous acrylic resin with a water content of 10 % (w/w), a product of Novozyme India Pvt. Ltd, Bangalore, India. All other chemicals and solvents were of analytical grade and obtained from Merck, India Ltd.

(ii) Method

The reaction was conducted with different molar ratios of capric acid and cetyl alcohol (1:1,1:2and 1:3) in a flat bottom flask and stirred at 200 rpm at different temperatures (45^{0}C to 65^{0}C) for different course of time with varying amounts of enzymes (4% to 12% w/w) of three different types of immobilized lipases (NS 435,RMIM and TLIM). This reaction was also performed in a complete solvent free environment and under vacuum (Vacuum pressure 28mmHg). The substrate molar ratio, temperature and enzyme loading of three different kinds of immobilized lipases (RMIM, NS 435 and TLIM) were the experimental parameters and optimized with a single reaction mixture that were studied one at a time and in due course followed by the subsequent reactions. The effect of three different lipases on the yield of the ester was also studied in this study. The entire reaction procedure has been provided in the schematic diagram **[Scheme 2]**.

The reaction period was monitored by determining the intermittent acid value of the product during the course of time. The ester formation was confirmed by thin layer chromatography (TLC), and the percent yield of ester and fatty acid composition of the ester was determined simultaneously. The ester was purified and stored for further uses.

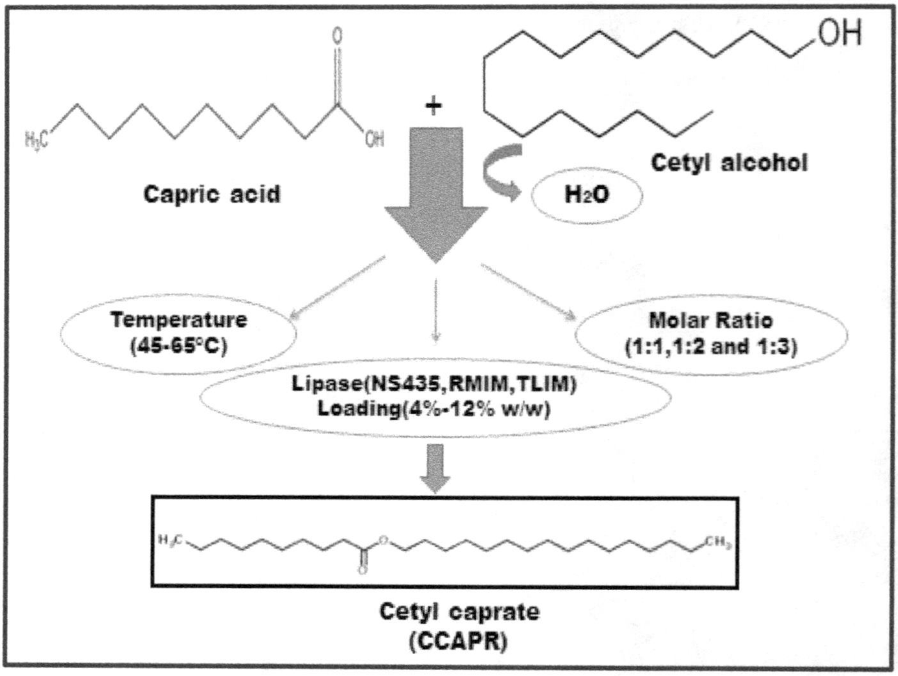

Scheme 2 .Reaction mechanism for the production of CCAPR from capric acid and cetyl alcohol

PHYSICOCHEMICAL CHARACTERIZATION OF EDIBLE PLASTICIZERS

Determination of acid value:

Acid value of the initial and intermittent reaction mixtures was determined by following standard method of AOCS [84] after separating the enzyme from the reaction mixture. About 2g of reaction mixture was weighed in a 250ml Erlenmeyer flask and about 25 ml of hot neutralized alcohol was added. Thereafter, the mixture was instantaneously titrated against 0.1 (N) NaOH by adding 0.5 ml of phenolphthalein indicator and the end point was identified with the appearance of light pink color which persisted for 30 seconds. From the decrease of acid value the yield of ester was calculated.

$$Acid\ value = \frac{56.1 \times Strength\ of\ NaOH \times Volume\ of\ NaOH}{Weight\ of\ the\ sample\ oil}$$

Qualitative analysis of esters by Thin Layer Chromatography (TLC):

Production of esters of caprylic acid or capric acid and cetyl alcohol were monitored by thin layer chromatography (TLC) by withdrawing samples from time to time and spotting the lipid mixtures on a Silica Gel G plate (0.2 mm thick). Here hexane-diethyl ether (90:10) was used as the developing solvent and different bands were visualized on iodine absorption and identified by measuring R_f value [85].

Analysis of fatty acid composition of pure caprylic acid, capric acid and its esters by GC:

The ester produced was first hydrolyzed with alkali and then the liberated free fatty acids were methylated with BF_3 in methanol and subjected to gas chromatographic (GC) analysis [86]. The GC (Agilent 6890 N) was fitted with a DB-Wax capillary column (30 m x 0.32 mm x 0.25 μm) and a flame ionization detector. The rate of N_2, H_2 and airflow was kept at 1, 30 and 300 ml/min, respectively. The inlet and detector temperatures were kept at 250^0C and the oven temperature was programmed to increase from 150^{0}C to 190^{0}C at a rate of 15^{0}C/min, then to hold for 5 min, and then to increase to 230^{0}C at a rate of 4^{0}C/min, and then to hold for 20 min.

FTIR Analysis:

Fourier transform infrared spectroscopy (FTIR) has been carried out. FTIR of the products was recorded by Perkin Elmer Spectrum GX spectrophotometer in the range 400–4000 cm^{-1}.

Physical parameters:

Physical parameters such as flash point, slip melting point, viscosity and density were determined by using ASTM methods [ASTM D56-05, ASTM D 93-80, ASTM D 445-04C2, ASTM D891-2].

ANTI-MICROBIAL ACTIVITY OF CETYL CAPRYLATE AND CETYL CAPRATE BY AGAR WELL DIFFUSION METHOD

(i) Materials

Test Samples for anti-microbial activity:

The prepared esters cetyl caprylate (CCAPL) and cetyl caprate (CCAPR)

Control:

Glycerol is a very well-known edible plasticizer in the food industry .Thus it was used as control in this test to compare the antimicrobial activity with the test samples.

Microorganism:

Escherichia coli (k8813) and *Staphylococcus aureus (ATCC 25923)*were provided by Institute of Microbial Technology, Chandigarh, India. Further the bacterial cultures were inoculated separately on nutrient agar slants and incubated at 37^0C for 24 h in incubator.

Media:

Nutrient agar media was used in this study for the growth of the microorganisms. This medium contained gram per litre (g/l) of distilled water Peptone-10.00,Beef extract-10.00 and Sodium Chloride (NaCl)-5,Agar-20.pH of the media 7.3.

(ii) Method

To study the antimicrobial activity of the plasticizers (CCAPL and CCAPR), agar well diffusion method was used to determine the zone of inhibition.

Agar well diffusion method:

25 ml of nutrient agar was poured on glass petriplates of same size and allowed to solidify. Agar surface of each plate was streaked by a sterile cotton swab with the reference bacterial strain. Agar plate was punched with a sterile cork borer of 4 mm size and 100 μL of each sample was poured with micropipette in the bore. The plates were allowed to standby for 30 min. The plates were incubated at 37^0C for 24 h.

PREPARATION AND PERFORMANCE EVALUATION OF THE CETYL CAPRYLATE AND CETYL CAPRATE ESTER PLASTICIZED CARBOXYMETHYL CELLULOSE (CMC) FILMS

(i)Materials

Carboxymethyl cellulose (CMC), with an average molecular weight of 41,000 g/mol (practical grade) was obtained from Merck (India). Glycerol and polysorbate 80 (Tween 80) were of analytical grade and were purchased from Merck (Darmstadt, India).

(ii) Method

Preparation of films:

Films were prepared as described by Ghanbarzadeh and Almasi [1] with some modifications. CMC (1 g) and varying amounts of glycerol were dissolved in a solvent mixture of 33 ml water and 66 ml ethanol at about 60^0C. After the addition of CCAPL and CCAPR separately in varying amounts and Tween 80 as emulsifier, the solutions were mixed and homogenized with an Ultra-Turrax T25 homogenizer at 24,000 rpm for 5 min. The abbreviations for the different samples prepared and their compositions are listed in the **Table 1**. After cooling and removing air bubbles, each of the samples (50 ml) was poured into an even surface Teflon mold (10–10 cm, 0.01 mm thick), dried at 60^0C, and then cast. The films obtained were thin, homogeneous, flexible and transparent.

Table 1.CMC films compositions with or without plasticizers

Samples	Sample Compositions				
	CMC (g)	**Glycerol (ml)**	**CCAPL (ml)**	**CCAPR (ml)**	**Tween 80 (% of esters)**
Pure CMC Films (Control)	1	0	0	0	0
CMC-GLY[a]	1	0.5	0	0	0
CMC-CCAPL1[b]	1	0.5	0.3	0	1
CMC-CCAPL2	1	0.5	0.6	0	1
CMC-CCAPL3	1	0.5	0.8	0	1
CMC-CCAPR1[c]	1	0.5	0	0.3	1
CMC-CCAPR2	1	0.5	0	0.6	1
CMC-CCAPR3	1	0.5	0	0.8	1

[a] CMC-GLY- Carboxymethyl cellulose + glycerol

[b] CMC-CCAPL- Carboxymethyl cellulose + cetyl caprylate ester

[c] CMC-CCAPR-Carboxymethyl cellulose + cetyl caprate ester

X-Ray diffraction:

In order to observe alterations of crystallinity of the films induced by CCAPL and CCAPR ester and glycerol as plasticizer addition, X-ray diffraction was performed on film samples with a P.W 3040/60model X-ray diffractometer PAN anaylitical (Netherland). The X-ray source was operated at 40 kV and 30 mA. Angular scans from5.02 to 99.97°(2Θ) were performed at a 2°C/min scan rate with a Cu source X-ray tube.

Scanning electron microscopy (SEM):

A thin layer of Au was sputter-coated with CMC film samples to observe the SEM images. Thereafter, the coated samples were observed by Zeiss EVO 18 Special Edition (Germany) with 15 kV accelerating voltage in secondary electron mode. The magnification was 10 kX in SEM images.

Mechanical properties:

A universal instrument of mechanical properties (Instron, 5565, USA) with a 50 N loaded cell was used for tensile strength, elongation at break and elastic modulus testing in a constant extension speed of 5 mm/min. Three film specimens, 8 × 0.5 cm strips, were cut from each of the film samples and were mounted between the grips of the machine. The initial grip separation and cross-head speed were set to 50 mm/min and 5 mm/min, respectively.

Thermo gravimetric analysis (TG):

TG was performed in an aluminum pan by TG-DTA thermal analyzer from Perkin Elmer, using software StareSW9.10. Flow rate of Nitrogen of 25 ml/min was used as the purging gas. The heating range of samples was from 20^0C to 200^0C with a rate of 10^0C/min.

Water vapor permeability:

Water vapor permeability of the CMC films with or without plasticizer (CCAPL, CCAPR and glycerol) was calculated in accordance with the modified ASTM E96-00 method. Film samples were sealed over a 60 mm circular opening of a permeable container with calcium chloride (0% RH inside the cell). Then the permeable containers were placed inside a desiccator containing the saturated sodium chloride solution (75% RH outside the cell), to create a 75% RH gradient across the film [87]. The water vapor transport was determined from the weight gain of the permeable container at an interval of every 24 h until the stable weight was attained. Water vapor transmission rate was calculated by using the following equation [88]:

$$Q = {WL}/{S}$$

where W is increase in the desiccant weight per 24 h, L is the film thickness (cm), S is the exposed surface area (cm^2) and Q is the water vapor transmission rate ($g/cm^2/24$ h).

Moisture uptake :

The CMC film samples (20 mm × 20 mm) were first conditioned using anhydrous $CaSO_4$ at 0% RH for 48 h. After weighing, they were conditioned at 25°C in a desiccator containing K_2SO_4 saturated solution [1]. This solution produces a relative humidity of 97%. The samples were removed from the desiccator after 48 h and weighed. The moisture uptake values of the samples were calculated as follows:

$$moisture\ uptake\% = \frac{W - W_0}{W_0} \times 100$$

Where, W is the weight of the film sample after exposure to 97% RH for 48 h, and W_0 is the initial weight of the sample. All measurements were performed in triplicate.

STATISTICAL ANALYSIS

All the analysis was done in triplicates (n = 3). All the data are represented as means with their standard errors (Mean ±SEM). Statistical comparisons among groups were performed using one way ANOVA. Level of significance $p < 0.05$ was considered by Tukey test (using Origin Pro 8 software).

RESULT AND DISCUSSION

PARAMETRIC DEPENDENCE FOR THE PRODUCTION OF CCAPL ESTER

Using the three distinct lipases (NS 435,RMIM and TLIM) the yield and characteristics of the esterified products were studied. At different temperature (45^{0}C to 65^{0}C), enzymatic esterification reaction with 1:1, 1:2, 1:3, molar ratios of caprylic acid and cetyl alcohol were performed in presence of these three enzymes for 0,1,2,3,4,5, and 6 h respectively.

Table 2 and Table 3 (vide Figure 1 a-I, b-I, c-I and a-II, b-II, c-II) ,Table 4 and Table 5 (vide Figure 2 a-I, b-I, c-I and a-II, b-II, c-II) and **Table 6 and Table 7 (vide Figure 3 a-I, b-I, c-I and a-II, b-II, c-II)** depicted the acid values and ester yields respectively at different temperature(45^0-65^{0}C) ,molar ratio (1:1,1:2 and 1:3) of caprylic acid and cetyl alcohol, and lipase loading (4-12%w/w) of the reaction were determined for three different lipases i., e NS 435 ,RMIM and TLIM respectively.

The progress of esterification reaction could easily be monitored by change in acid value of the reaction mixture. With the formation of ester, the acid value should decrease due to the loss of free fatty acid from the reaction mixture. Figure 1 (a-I) and Figure 1(a-II) showed the change in acid value and yield of the ester with different temperature respectively, keeping the molar ratio (1:1 caprylic acid: cetyl alcohol) and NS 435 lipase loading (10% w/w) constant. From the Figure 1(a-I) acid values of 1:1 molar ratio of caprylic acid and cetyl alcohol at different temperatures ranging from 45^{0}C to 65^{0}C were determined at different time interval (0 to 6 h) using immobilized NS 435 as biocatalyst (10% w/w).In case of 65^{0}C temperature, Figure 1(a-I) showed the highest decrease of acid value in comparison to other reaction temperatures (e.g. 45^{0}C, 50^{0}C, 55^{0}C and 60^{0}C). When the acid values of different temperatures were compared among each other, Figure 1(a-I) showed that the acid value at 65^{0}C was significantly ($p<0.05$) decreased in comparison to 45^{0}C, 50^{0}C and 55^{0}C. But acid value was not significantly ($p<0.05$) decreased in case of 60^{0}C when it was compared with

65^0C.Though the highest decrease in acid value was observed, therefore, from the Figure 1(a-I) a significant ($p<0.05$) decrease of acid value was obtained at 65^0C reaction temperature of caprylic acid and cetyl alcohol at 1:1 molar ratio using 10% (w/w) immobilized NS435 as biocatalyst. Figure 1 (a-II) depicted the yield of ester at different temperature (45^0C to 65^0C) and at different time interval (0 to 6 h), keeping the molar ratio (1:1) and lipase loading (10% w/w) constant. At 65^0C temperature, Figure 1(a-II) showed the highest increase of ester yield in comparison to other reaction temperature (e.g. 45^0C,50^0C, 55^0C and 60^0C).When the ester yield at different temperatures were compared among each other, Figure 1(a-II) showed that the ester yield at 65^0C was significantly($p<0.05$) increased in comparison to 45^0C,50^0C and 60^0C. But the ester yield was not significantly ($p<0.05$) increased in case of 60^0C when it was compared with 65^0C. With the increase in temperature from 45^0C up to 65^0C, there was an increase in the ester yield.

When immobilized RMIM and TLIM was applied respectively instead of NS435 enzyme, keeping the other reaction parameters same as in the previous case, the acid value at the 6h was found to be the least for 65^0C at 1:1 molar ratios and , as observed in Figure 1 (b-I and c-I). Figure 1 (b-II and c-II) depicted the ester yield of the reaction catalyzed by RMIM and TLIM respectively and maximum yield was obtained in 6h significantly ($p<0.05$). Maximum ester yield was obtained with the NS435 catalyzed reaction, when compared with yield of immobilized RMIM and TLIM catalyzed reaction.

This part of the study was concerned with the optimization of reaction temperature for conducting the reaction procedure of caprylic acid and cetyl alcohol. From Figure 1 (a-I,a-II,b-I,b-II and c-I,c-II), it can be stated that 65^0C reaction temperature was optimal for lipase catalysis and NS435 lipase performed better in comparison to other lipases. This phenomenon might be caused due to the random selectivity nature of the NS 435 lipase.

Table 2.Changes in Acid value at different reaction temperature (45^0 to 65^{0}C) at a constant molar ratio (1:1) of caprylic acid and cetyl alcohol using NS 435, RMIM and TLIM as biocatalyst (10%w/w) respectively at different time intervals (0 - 6 h)

Lipase	Time(h) ↓	Acid Value				
		Temperature(^{0}C)				
		45	**50**	**55**	**60**	**65**
NS435	0	350.2±0.20	349.02±0.10	350.01±0.10	350.09±0.10	351.3±0.10
	1	240.08±0.02	209.03±0.03	180±0.03	156.08±0.15	146.6±0.02
	2	130.06±0.10	110.08±0.05	95.08±0.06	85.09±0.03	70.01±0.05
	3	82.08±0.05	69.56±0.11	50.08±0.11	38.09±0.05	28.53±0.12
	4	55.09±0.03	38.56±0.04	24.79±0.03	17.89±0.05	10.16±0.15
	5	38.9±0.105	27.06±0.03	18.09±0.03	10.89±0.07	5.16±0.15
	6	30.09±0.10	20.06±0.10	13.6±0.10	5.09±0.10	**1.28±0.12***
Lipase	**Time(h) ↓**	**45**	**50**	**55**	**60**	**65**
RMIM	0	350.2 ±0.20	349.02±0.10	350.01±0.10	349.6 ±0.10	351.3±0.20
	1	230.08±0.02	209.03±0.03	190.06±0.03	175.6±0.03	146.6±0.02
	2	140.06±0.10	120.08±0.05	100.08±0.05	89.66±0.04	71.06±0.10
	3	89.08±0.05	72.56±0.11	65.08±0.11	49.66±0.11	32.53±0.105
	4	65.09±0.02	48.56±0.02	39.79±0.03	28.7± 0.02	15.16±0.05
	5	43.9±0.11	37.06±0.03	28.09±0.03	17.36±0.03	8.16±0.02
	6	35.09±0.11	27.06±0.10	17.6±0.10	8.89±0.12	3.28±0.10
Lipase	**Time(h) ↓**	**45**	**50**	**55**	**60**	**65**
TLIM	0	350.2±0.20	349.02±0.10	350.01±0.10	349.6±0.10	351±0.10
	1	230.08±0.02	219.03±0.03	205.06±0.03	195.6±0.04	181.6±0.03
	2	145.06±0.10	130.08±0.05	120.08±0.06	110.6±0.04	99.33±0.05
	3	95.08±0.05	82.56±0.11	75.08±0.11	65.6±0.11	49.53±0.11
	4	72.09±0.02	58.56±0.03	49.79±0.05	42.7±0.03	30.76±0.05
	5	51.9±0.11	40.06±0.03	34.09±0.03	28.36±0.03	15.6±0.03
	6	42.09±0.10	32.06±0.10	24.6±0.10	15.90±0.10	7.47±0.10

*Values are represented as mean ± SEM. n=3, ($p<0.05$)

Table 3 .Yield of ester at different reaction temperature (45°C to 65°C) using NS 435,RMIM and TLIM as biocatalyst (10% w/w) respectively at a constant molar ratio (1:1) of caprylic acid and cetyl alcohol at a different reaction time (0-6h)

Lipase	Yield of Ester (%w/w)					
	Temperature(°C)					
	Time(h) ↓	45	50	55	60	65
NS435	0	0	0	0	0	0
	1	31.42±0.02	40.29±0.03	48.58±0.03	55.41±0.15	58.12±0.02
	2	62.84±0.1	68.55±0.04	72.84±0.05	75.69±0.03	80.00±0.05
	3	76.55±0.05	80.13±0.11	85.69±0.11	89.11±0.05	91.84±0.12
	4	84.26±0.03	88.98±0.02	92.91±0.03	94.88±0.05	97.09±0.15
	5	88.88±0.11	92.27±0.03	94.83± 0.03	96.88±0.07	98.52±0.15
	6	91.40±0.10	94.27±0.10	96.11±0.10	98.54±0.10	**99.63±0.12***
Lipase	Time(h) ↓	45	50	55	60	65
RMIM	0	0	0	0	0	0
	1	34.27±0.02	40.30±0.03	45.71±0.04	49.84±0.02	58.12±0.02
	2	59.99±0.10	65.70±0.05	71.41±0.06	74.38±0.04	79.71±0.10
	3	74.55±0.05	79.27±0.11	81.41±0.11	85.81±0.11	90.70±0.05
	4	81.40±0.02	86.13±0.03	88.63±0.03	91.80±0.025	95.66±0.02
	5	87.46±0.11	89.41±0.03	91.97±0.03	95.04±0.03	97.67±0.11
	6	89.97±0.10	92.27±0.10	94.97±0.10	97.46±0.10	99.06±0.10
Lipase	Time(h) ↓	45	50	55	60	65
TLIM	0	0	0	0	0	0
	1	34.27±0.02	37.43±0.02	41.42±0.03	44.12±0.02	48.12±0.02
	2	58.56±0.10	62.84±0.10	65.70±0.05	68.40±0.10	71.62±0.05
	3	72.84±0.05	76.41±0.05	78.55±0.11	81.26±0.05	85.85±0.03
	4	79.40±0.02	83.27±0.03	85.77±0.03	87.80±0.04	91.21±0.11
	5	85.17±0.10	88.55±0.11	90.26±0.03	91.89±0.11	95.54±0.10
	6	87.97±0.11	90.84±0.10	92.97±0.10	95.46±0.10	97.86±0.10

*Values are represented as mean ± SEM. n=3, ($p<0.05$)

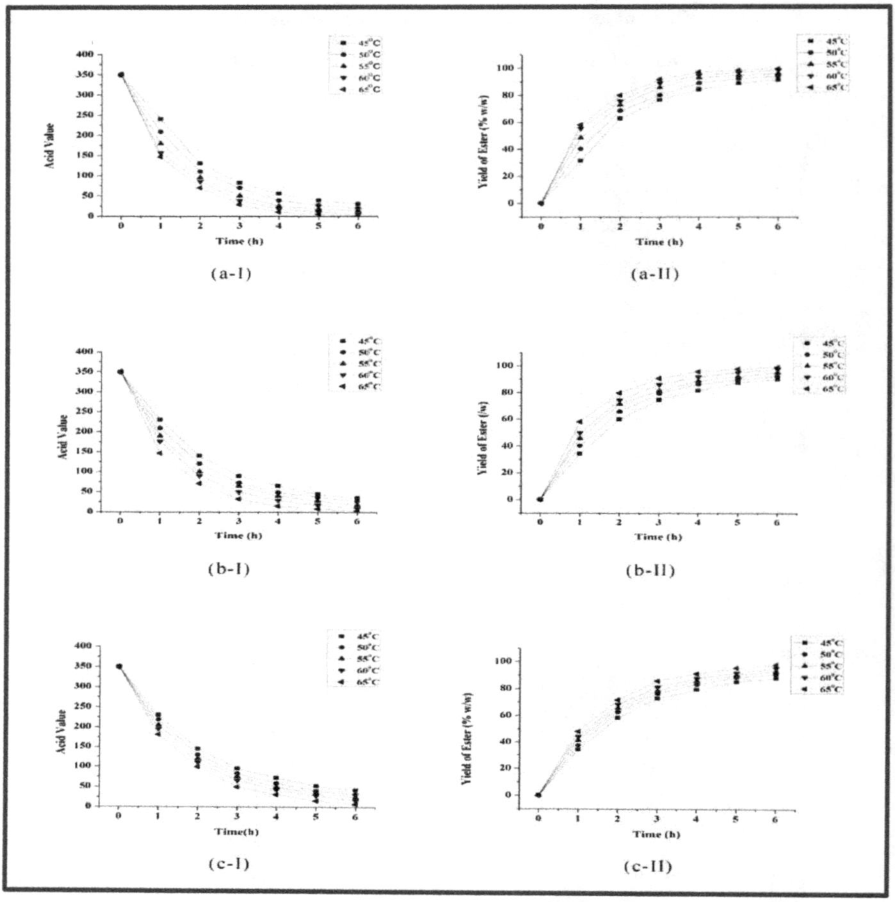

Figure 1. (a-I, b-I and c-I) Changes in Acid value at different reaction temperature (45 to 65^{0}C) at a constant molar ratio (1:1) of caprylic acid and cetyl alcohol using NS 435,RMIM and TLIM as biocatalyst (10%w/w) respectively at different time intervals (0-6 h); and(a-II,b-II and c-II)Yield of ester at different reaction temperature (45^{0}C to 65^{0}C) using NS 435,RMIM and TLIM as biocatalyst (10% w/w) respectively at a constant molar ratio (1:1) of caprylic acid and cetyl alcohol at a different reaction time (0-6h). Values are represented as mean ± SEM. n=3, ($p<0.05$)

Table 4 and Table 5 (vide Figure 2 a-I,b-I,c-I and a-II,b-II,c-II)depicted the change in acid value and yield of the ester respectively with different molar ratio of caprylic acid and cetyl alcohol (1:1 ,1:2 and 1:3), keeping both the temperature (65^{0}C) and lipase loading (10% w/w) constant. From Figure 2 (a-I), the acid values at 0 h to 6 h for 1:1, 1:2 and 1:3 molar ratios of caprylic acid and cetyl alcohol were determined using NS 435 as biocatalyst (10% w/w) at 65^{0}C.In case of 1:1molar ratio of caprylic acid and cetyl alcohol, Figure 2(a-I) showed that there was a decrease in acid value with time and the highest decrease in acid value was obtained at the 6 h of the reaction. This pattern of the reaction was also observed in other molar ratios (e.g.1:2 and 1:3) of caprylic acid and cetyl alcohol at 0 to 6 h reaction time. Same pattern was maintained in case of Figure 2 (b-I and c-I) where RMIM and TLIM were used respectively as catalysts. But when the performance of these three catalysts were compared with each other, immobilized NS435 showed the highest decrease in acid value at 1:1 molar ratio of caprylic acid and cetyl alcohol [Figure 2(a-I)].This occurrence might be caused due to the selection of random lipase NS 435 in comparison to other lipases (RMIM and TLIM) which are 1, 3 specific in nature.

At the same time, Figure 2 (a-II, b-II and c-II) described the ester yield of ester with time (0 h to 6 h) at different molar ratios (1:1, 1:2 and 1:3) of caprylic acid and cetyl alcohol at constant temperature (65^{0}C) and constant lipase loading (10% w/w) with the corresponding lipases (NS 435,RMIM and TLIM).Figure 2 (a-II) showed highest ester yield at 1:1 molar ratio of caprylic acid and cetyl alcohol compared to the molar ratios of 1:2 and 1:3 , when immobilized NS 435 lipase was used as catalyst. But when the reactions were conducted by RMIM (Figure 2 b-II) (10% w/w) and TLIM (Figure 2 c-II) (10% w/w) as catalysts respectively ,the yields were significantly ($p<0.05$) lower than the yield of the immobilized NS 435 lipase (10% w/w) at 1:1 molar ratio of caprylic acid and cetyl alcohol, at a temperature of 65^{0}C. Each type of lipase catalyzed reactions showed same pattern where 1:1

molar ratio of caprylic acid and cetyl alcohol showed highest decrease in acid value and maximum yield of ester at 65°C and 10% (w/w) lipase concentration.

Hence from the Figure 2 (a-I, a-II, b-I, b-II and c-I, c-II), it can be stated that optimum molar ratio for the reaction of caprylic acid and cetyl alcohol was 1:1 which was catalyzed by immobilized NS 435 lipase at 65°C.

Table 4 .Changes in Acid value at different reaction molar ratio (1:1,1:2 and 1:3) of caprylic acid and cetyl alcohol using NS 435,RMIM and TLIM as biocatalyst at a constant lipase loading (10%w/w) and at a constant reaction temperature (65°C) at different time intervals (0-6 h)

Lipase	**Time (h)** ↓	**Acid Value**		
		Molar Ratio(Caprylic acid and Cetyl alcohol)		
		1:1	**1:2**	**1:3**
NS435	0	351.30±0.10	350.09±0.10	350.04±0.10
	1	146.60±0.05	155.09±0.15	165.02±0.10
	2	70.01±0.05	79.02±0.07	85.02±0.05
	3	28.53±0.20	35.06±0.10	45.02±0.10
	4	10.16±0.13	17.89±0.12	25.05±0.13
	5	5.16±0.15	14.09±0.05	22.01±0.04
	6	**1.28±0.15***	10.06±0.06	18.01±0.06
Lipase	**Time (h)** ↓	**1:1**	**1:2**	**1:3**
RMIM	0	351.30±0.10	350.09±0.10	350.04±0.10
	1	146.6±0.03	165.09±0.10	185.02±0.15
	2	71.01±0.05	89.02±0.05	100.02±0.05
	3	32.53±0.12	45.06±0.10	55.02±0.10
	4	15.16±0.15	27.89±0.12	35.05±0.12
	5	8.16±0.15	17.09±0.05	27.01±0.05
	6	5.89±0.12	14.06±0.06	22.01±0.06
Lipase	**Time (h)** ↓	**1:1**	**1:2**	**1:3**
TLIM	0	351.01±0.10	350.09 ±0.10	350.04±0.10
	1	181.6±0.03	195.09±0.15	205.02±0.15
	2	99.33±0.05	109.02±0.05	120.02±0.05
	3	49.53±0.12	65.06±0.10	75.02±0.10
	4	30.76±0.15	42.89±0.12	55.05±0.12
	5	15.6±0.15	30.09±0.05	40.01±0.05
	6	7.47±0.12	18.06±0.06	27.01±0.06

*Values are represented as mean ± SEM. n=3, ($p<0.05$)

Table 5 .Yield of ester at different molar ratio(1:1,1:2 and 1:3) of caprylic acid and cetyl alcohol using NS 435,RMIM and TLIM as biocatalyst in a constant lipase loading (10% w/w) at a constant reaction temperature (65^{0}C) at different reaction time intervals (0-6h)

Lipase	Time ↓	Yield of Ester (%w/w)		
		Molar Ratio(Caprylic acid and Cetyl alcohol)		
		1:1	1:2	1:3
NS435	0	0	0	0
	1	58.12±0.03	55.69±0.11	52.86±0.15
	2	80.00±0.05	77.42±0.05	75.71±0.05
	3	91.84±0.12	89.98±0.10	87.14±0.10
	4	97.09±0.15	94.88±0.12	92.84±0.12
	5	98.52±0.15	95.97±0.05	93.71±0.05
	6	**99.63±0.15***	97.12±0.06	94.85±0.06
Lipase	**Time (h) ↓**	**1:1**	**1:2**	**1:3**
RMIM	0	0	0	0
	1	58.12±0.03	52.84±0.15	47.15±0.12
	2	79.71±0.05	74.57±0.15	71.43±0.05
	3	90.70±0.11	87.12±0.05	84.28±0.10
	4	95.66±0.14	92.03±0.10	89.98±0.12
	5	97.66±0.16	95.11±0.12	92.28±0.05
	6	98.31±0.12	95.98±0.05	93.71±0.06
Lipase	**Time (h) ↓**	**1:1**	**1:2**	**1:3**
TLIM	0	0	0	0
	1	48.12±0.03	44.27±0.18	41.43±0.15
	2	71.62±0.01	68.85±0.05	65.71±0.06
	3	85.85±0.12	81.41±0.11	78.57±0.10
	4	91.21±0.15	87.74±0.12	84.27±0.14
	5	95.54±0.15	91.40±0.05	88.57±0.05
	6	97.86±0.12	94.84±0.07	92.28±0.06

*Values are represented as mean ± SEM. n=3, ($p<0.05$)

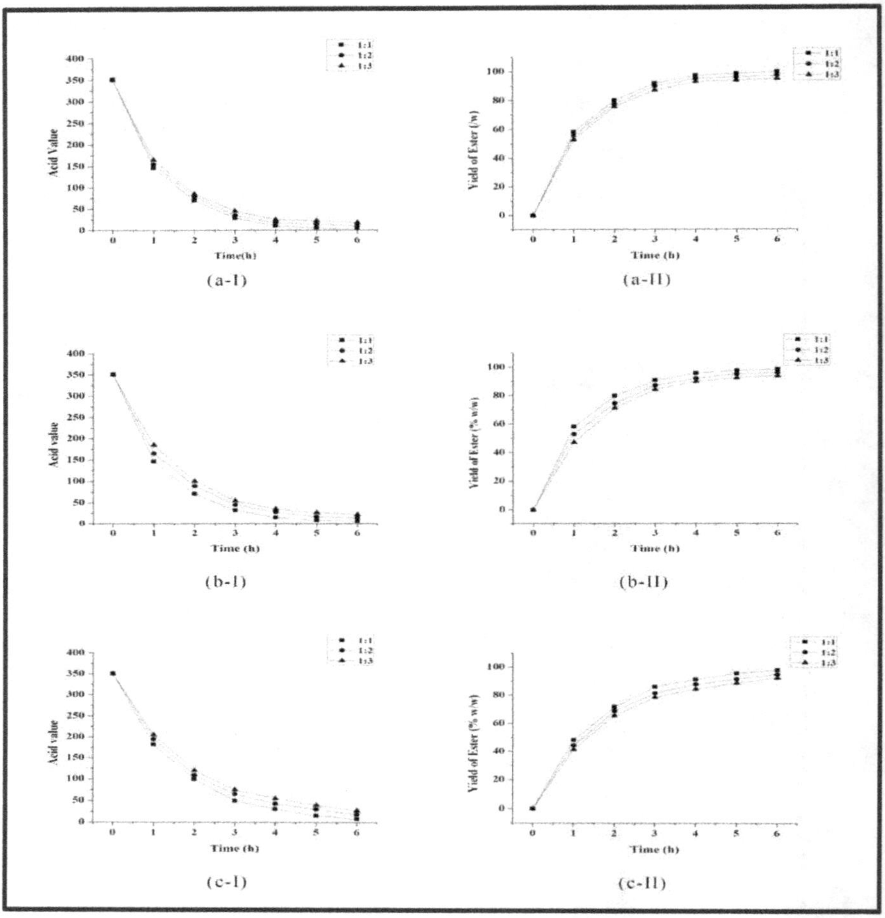

Figure 2. (a-I, b-I and c-I) Changes in Acid value at different reaction molar ratio (1:1,1:2 and 1:3) of caprylic acid and cetyl alchol using NS 435,RMIM and TLIM as biocatalyst at a constant lipase loading (10%w/w) and at a constant reaction temperature (65^{0}C) at different time intervals (0-6 h); and(a-II, b-II and c-II) Yield of ester at different molar ratio(1:1,1:2 and 1:3) of caprylic acid and cetyl alcohol using NS 435, RMIM and TLIM as biocatalyst in a constant lipase loading (10% w/w) at a constant reaction temperature (65^{0}C) at different reaction time (0-6h).Values are represented as mean ± SEM. n=3,($p<0.05$)

From **Table 6 and Table 7(vide Figure 3 a-I, a-II, b-I, b-II and c-I, c-II)**,the effect of immobilized lipase concentration on the reaction was determined with different immobilized lipases (NS435,RMIM and TLIM) at constant substrate molar ratio of caprylic acid and cetyl alcohol (1:1) and constant temperature of reaction (65^{0}C). On the basis of total substrate weight 4%, 6%, 8%, 10% and 12% (w/w) immobilized lipase loading were considered. In case of the reaction catalyzed by immobilized NS 435 lipase (Figure 3 a-I and a-II), it was observed that, with the increase in immobilized lipase concentration the acid value was decreasing with corresponding ester yield. Highest decrease in acid value and maximum yield of ester was obtained at 6h of reaction time at 10% (w/w) NS 435 lipase concentration. But RMIM (Figure 3 b-I and b-II) and TLIM (Figure 3 c-I and c-II) catalyzed reaction showed significantly lower acid value and ester yield in comparison to immobilized NS 435 lipase. It might be due to the random selectivity of the immobilized NS 435 lipase. But in each set of the reaction catalyzed by these three kinds of lipases, there was the same trend where 10% (w/w) lipase concentration showed invariably highest acid value drop and maximum ester yield.

Hence from the Figure 3 (a-I,a-II,b-I,b-II and c-I,c-II),it can be said that the performance of the NS435 was better in comparison to other lipases and the lipase load for the reaction of caprylic acid and cetyl alcohol was 10% (w/w).

Therefore, the standardized reaction parameter for maximum yield of CCAPL ester , the lipase being immobilized NS 435, was found at 6 h with the reactant molar ratio of 1:1 and 65^{0}C reaction temperature.

Table 6 .Changes in Acid value at different lipase loading (4,6,8,10 and 12% w/w) using NS 435, RMIM and TLIM as biocatalyst at a constant molar ratio (1:1) of caprylic acid and cetyl alcohol at constant temperature (65°C) with different time intervals (0- 6 h)

Lipase	Time(h) ↓	Acid Value Lipase loading(%w/w) 4	6	8	10	12
NS435	0	350.09±0.10	350.1±0.10	350.09±0.10	351.3±0.10	350.1±0.10
	1	219.09±0.15	199.8±0.15	175.4±0.15	146.6±0.15	159.09±0.15
	2	149.08±0.05	120.06±0.05	105.08±0.05	70.02±0.05	88.05±0.05
	3	98.8±0.20	78.01±0.20	61.05±0.20	28.53±0.20	40.4±0.20
	4	70.09±0.05	52.01±0.05	39.08±0.05	10.16±0.05	24.9±0.05
	5	40.98±0.02	25.09±0.03	17.67±0.02	5.16±0.03	13.4±0.02
	6	35.09±0.15	22.08±0.15	12.5±0.15	**2.28±0.15***	9.5±0.15
Lipase	**Time(h) ↓**	**4**	**6**	**8**	**10**	**12**
RMIM	0	350.09±0.10	350.1±0.10	350.09±0.10	349.6±0.10	350.1±0.10
	1	220.09±0.15	201.8±0.15	185.4±0.15	156.6±0.15	169.09±0.15
	2	159.08±0.05	140.06±0.05	125.08±0.05	89.66±0.05	105.05±0.05
	3	106.80±0.2	88.01±0.20	72.05±0.20	43.66±0.20	59.4±0.20
	4	75.09±0.03	62.01±0.02	47.08±0.03	25.06±0.03	36.9±0.02
	5	50.98±0.15	40.09±0.15	29.67±0.15	12.36±0.15	18.4±0.15
	6	40.09±0.05	28.08±0.05	16.5±0.05	5.89±0.05	12.08±0.05
Lipase	**Time(h) ↓**	**4**	**6**	**8**	**10**	**12**
TLIM	0	350.09±0.10	350.10±0.10	350.09±0.12	351.01±0.10	350.10±0.10
	1	209.09±0.15	199.8±0.15	190.4±0.1	161.6±0.15	179.09±0.15
	2	160.08±0.20	150.06±0.20	135.08±0.20	99.33±0.20	115.05±0.20
	3	110.80±0.05	100.01±0.05	82.05±0.06	49.53±0.05	69.4±0.05
	4	82.09±0.03	72.01±0.02	57.08±0.12	30.76±0.03	46.9±0.02
	5	65.98±0.15	55.09±0.15	39.67±0.10	15.6±0.15	28.4±0.15
	6	45.09±0.10	35.08±0.10	25.5±0.20	7.47±0.10	18.08±0.10

*Values are represented as mean ± SEM. n=3, ($p<0.05$)

Table 7.Yield of ester at different lipase loading (4,6,8,10 and 12% w/w) using NS 435,RMIM and TLIM as biocatalyst respectively at a constant molar ratio (1:1) of caprylic acid and cetyl alcohol and at constant reaction temperature (65^0C) with different reaction time (0-6h) intervals

Lipase	Time(h) ↓	Yield of Ester (%w/w)				
		Lipase loading(%w/w)				
		4	6	8	10	12
NS435	0	0	0	0	0	0
	1	42.27±0.05	46.90±0.03	50.36±0.15	57.35±0.02	54.19±0.02
	2	61.51±0.11	64.32±0.05	67.28±0.20	73.06±0.10	71.12±0.10
	3	71.13±0.02	75.04±0.12	77.66±0.05	83.71±0.05	80.76±0.05
	4	78.58±0.03	82.66±0.15	85.21±0.02	90.04±0.02	87.30±0.03
	5	84.58±0.10	88.81±0.15	91.51±0.15	95.81±0.11	93.44±0.12
	6	90.35±0.10	91.25±0.12	94.20±0.10	**99.4±0.10***	97.68±0.10
Lipase	**Time(h) ↓**	**4**	**6**	**8**	**10**	**12**
RMIM	0	0	0	0	0	0
	1	34.56±0.15	42.35±0.10	47.04±0.15	55.26±0.15	51.70±0.10
	2	54.56±0.20	59.99±0.20	64.27±0.06	74.38±0.20	69.99±0.20
	3	69.49±0.05	74.86±0.06	79.41±0.03	87.52±0.05	83.03±0.06
	4	77.12±0.03	82.28±0.12	86.55±0.10	92.84±0.02	89.45±0.12
	5	85.43±0.15	88.54±0.10	91.52±0.11	96.46±0.15	94.74±0.10
	6	87.12±0.1	91.97±0.20	95.28±0.10	98.31±0.10	96.54±0.20
Lipase	**Time(h) ↓**	**4**	**6**	**8**	**10**	**12**
TLIM	0	0	0	0	0	0
	1	40.27±0.15	42.92±0.15	45.61±0.10	53.84±0.15	48.84±0.15
	2	54.27±0.20	57.13±0.20	61.41±0.20	71.62±0.20	67.13±0.20
	3	68.35±0.05	71.43±0.05	76.56±0.06	85.85±0.05	80.17±0.05
	4	76.55±0.02	79.43±0.03	83.69±0.12	91.21±0.02	86.60±0.04
	5	81.15±0.15	84.26±0.15	88.66±0.10	95.54±0.15	91.88±0.15
	6	87.12±0.10	89.98±0.10	92.71±0.20	97.86±0.10	94.83±0.10

*Values are represented as mean ± SEM. n=3, ($p<0.05$)

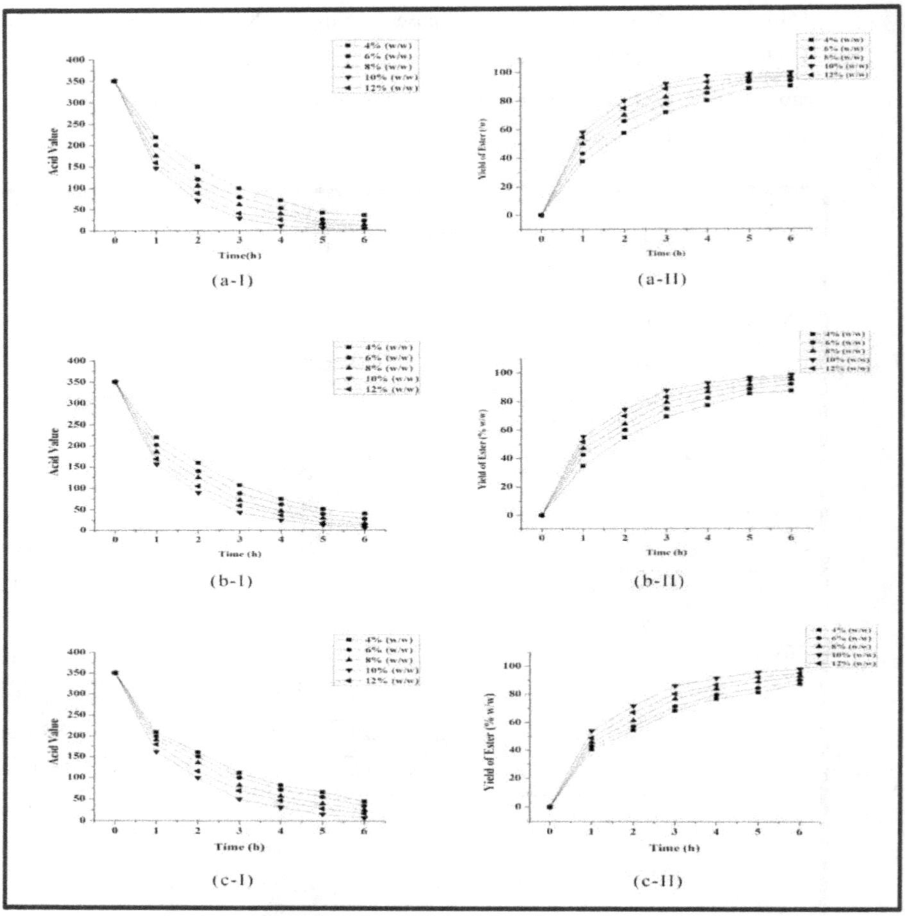

Figure 3. (a-I, b-I and c-I) Changes in Acid value at different lipase loading (4,6,8,10 and 12% w/w) using NS 435, RMIM and TLIM as biocatalyst at a constant molar ratio (1:1) of caprylic acid and cetyl alcohol at constant temperature (65°C) with different time intervals (0-6h) ; and(a-II ,b-II and c-II) Yield of ester at different lipase loading (4,6,8,10 and 12% w/w) using NS 435,RMIM and TLIM as biocatalyst respectively at a constant molar ratio (1:1) of caprylic acid and cetyl alcohol and at constant reaction temperature (65°C) with different reaction time (0-6h) intervals. Values are represented as mean ± SEM, n=3, ($p<0.05$)

FTIR analysis of pure caprylic acid and CCAPL ester:

FTIR spectra of pure caprylic acid ,cetyl alcohol and CCAPL ester was presented in the **Table 8 (vide Figure 4 a, b and c)** .The existence of cyclic dimeric structure of caprylic acid is characterized by a very broad OH stretching band centering near 3000 cm^{-1} and a medium intensity band at 1107 cm^{-1} (Figure 4 a) due to the out of plane O…H…O hydrogen deformation [89]. C−H stretching (2500-3000 cm^{-1}), C=O stretching vibration (1712 cm^{-1}) , and C−O stretching vibrations was found at several intensities in case of the FTIR spectra of caprylic acid (Figure 4 a).Figure 4(b) described the FTIR spectra of cetyl alcohol. O−H stretching vibration(3325.1 cm^{-1}),C−H stretching vibration (2918.2 cm^{-1}) and C−O stretching (1463 cm^{-1},1064.3 cm^{-1}, and 719.7 cm^{-1}) was found to be observed in case of the FTIR spectral analysis of cetyl alcohol. Figure 4 c depicted the FTIR spectral analysis of CCAPL ester. The O−H stretching vibration (3300 cm^{-1}),C−H stretching vibration (2969.5 cm^{-1}), C=O stretching vibration (1712 cm^{-1}) and C−O stretching vibration (1468.9) were observed at sharp intensities which confirmed the CCAPL ester formation.

Table 8.FTIR peak assignment for cetyl alcohol, caprylic acid, CCAPL ester

Group of class	Range(cm^{-1})		
	Cetyl Alcohol	Caprylic Acid	CCAPL ester
O-H and C-H stretching vibration	2910-2850	2900-3000	3300-2969.5
O-H bending	1100	948-1419	940-1403
C-O stretching vibrations	1463- 719.7	1296	1468.9
C=O stretching vibration	-	1721	1712
C-H rocking vibration	1470-717	-	1465-709
C-H bending	-	-	-

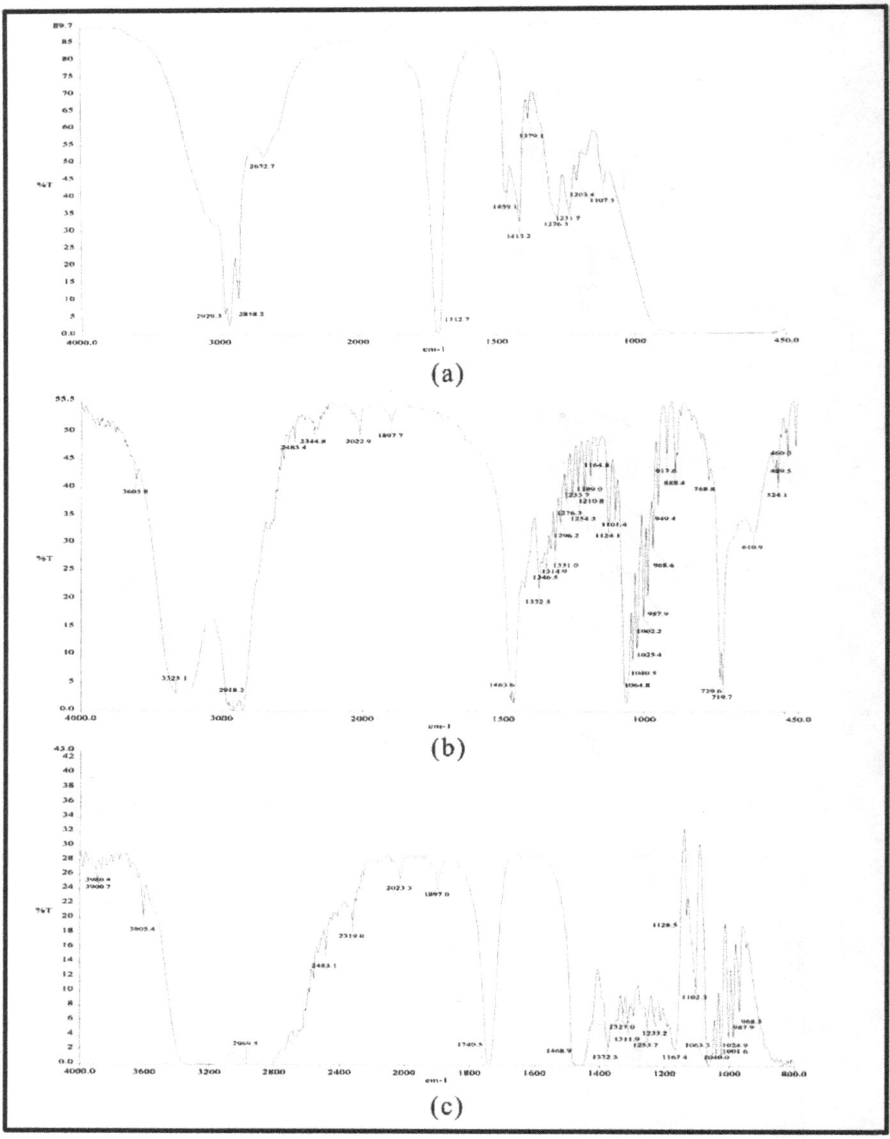

Figure 4. FTIR spectra of (a) Pure caprylic acid, (b) Cetyl alcohol and (c) CCAPL ester

Physical characterization of caprylic acid and CCAPL ester

The physical characterization of the caprylic acid and CCAPL ester were listed in the **Table 9**. The physical properties of the caprylic acid and CCAPL ester obtained were comparable with the results obtained by other researchers. The flash point was obtained above the 280°C. The viscosity of caprylic acid is 5.02 cP at 50°C temperature, whereas for ester of caprylic acid and cetyl alcohol, it was 4.82 at 50°C. The density of the caprylic acid and CCAPL ester was almost similar.The slip melting point of CCAPL ester was slightly lower than caprylic acid.

Table 9. Physical characterization of caprylic acid and CCAPL ester

Properties	Values	
	Caprylic acid	**CCAPL ester**
Flash Point (°C)	130±0.05	Above 280
Viscosity(cP at 50°C)	5.02 ±0.15	4.82±0.03
Density (g/cm^3) at 30°C	0.904±0.015	0.844±0.10

*Values are represented as mean ± SEM. n=3, ($p<0.05$)

PARAMETRIC DEPENDENCE FOR THE PRODUCTION OF CCAPR ESTER

The same reaction parameters were set for capric acid and cetyl alcohol. **From Table 10 and Table 11 (vide Figure 5 a-I, a-II, b-I, b-II and c-I, c-II)**, the standard reaction parameters were found to be 6 h reaction time, 1:1 molar ratio of capric acid and cetyl alcohol, and at 65°C with NS 435 immobilized lipase used.

Table 12 and Table 13(vide Figure 6 a-I, a-II, b-I, b-II and c-I, c-II) described the changes in acid value at different reaction temperature (45°,50°,55°,60° and 65°C) at a constant molar ratio (1:1) of capric acid and cetyl alcohol using NS 435,RMIM and TLIM. When the acid values of all the molar ratios (1:1, 1:2 and 1:3) at 0 to 6 h time interval was plotted using the statistical software (Figure 6a-I,a-II,b-I,b-II and c-I, c-II), significant ($p<0.05$) drop of acid value and maximum yield of ester was observed at the 6 h for the 1:1 molar ratio of capric acid and cetyl alcohol catalyzed by NS435 immobilized lipase at 65^{0}C.

Table 10 .Changes in Acid value at different reaction temperature (45^0 to 65^{0}C) at a constant molar ratio (1:1) of capric acid and cetyl alcohol using NS 435, RMIM and TLIM as biocatalyst (10%w/w) respectively at different time intervals (0 - 6 h)

Lipase	Time(h) ↓	Acid Value Temperature(^{0}C) 45	50	55	60	65
NS435	0	260.01±0.20	260.50±0.10	260.09±0.10	260.05±0.10	260.56±0.20
	1	150.08±0.02	140.89±0.03	130.56±0.04	120.09±0.03	110.89±0.02
	2	110.06±0.10	98.09±0.06	91.87±0.05	79.89±0.04	70.03±0.10
	3	86.2±0.05	76.82±0.11	69.09±0.11	58.45±0.11	42.34±0.05
	4	65.09±0.02	55.06±0.03	47.09±0.02	35.67±0.03	25.89±0.05
	5	45.08±0.15	35.09±0.03	28.09±0.03	18.9±0.03	10.89±0.11
	6	26.09±0.10	20.09±0.10	15.08±0.10	7.01±0.10	**1.56±0.10***
Lipase	**Time(h) ↓**	**45**	**50**	**55**	**60**	**65**
RMIM	0	260.01±0.20	260.50±0.10	260.09±0.10	260.05±0.10	260.56±0.20
	1	165.08±0.02	148.89±0.03	136.56±0.03	128.09±0.03	118.89±0.02
	2	110.06±0.10	100.09±0.05	92.87±0.04	83.09±0.04	75.03±0.10
	3	89.07±0.05	80.82±0.11	69.09±0.11	62.45±0.11	48.34±0.05
	4	68.09±0.03	60.06±0.03	49.09±0.02	39.67±0.02	32.89±0.03
	5	49.08±0.11	40.09±0.03	32.09±0.03	21.90±0.03	15.89±0.11
	6	35.09±0.10	26.09±0.10	19.08±0.10	11.01±0.10	6.56±0.10
Lipase	**Time(h) ↓**	**45**	**50**	**55**	**60**	**65**
TLIM	0	260.02±0.20	260.50±0.10	260.09±0.10	260.05±0.10	260.56±0.20
	1	170.08±0.02	158.89±0.03	146.56±0.03	138.09±0.03	128.89±0.02
	2	120.06±0.10	110.09±0.05	102.87±0.06	93.09±0.05	85.03±0.10
	3	99.07±0.05	90.82±0.10	79.09±0.11	72.45±0.11	58.34±0.05
	4	78.09±0.03	70.06±0.02	59.09±0.02	49.67±0.02	39.89±0.05
	5	59.08±0.11	50.09±0.03	42.09±0.03	31.9±0.03	25.89±0.11
	6	45.09±0.10	36.09±0.11	29.08±0.10	21.01±0.10	12.56±0.10

*Values are represented as mean ± SEM. n=3, ($p<0.05$)

Table 11.Yield of ester at different reaction temperature (45°C to 65°C) using NS 435,RMIM and TLIM as biocatalyst (10% w/w) respectively at a constant molar ratio (1:1) of capric acid and cetyl alcohol at a different reaction time intervals (0-6h)

Lipase	Time(h) ↓	Yield of Ester(%w/w) Temperature(°C)				
		45	50	55	60	65
NS435	0	0	0	0	0	0
	1	42.27±0.02	45.81±0.03	49.78±0.02	53.81±0.03	57.35±0.01
	2	57.66±0.10	62.27±0.05	64.66±0.06	69.27±0.05	73.06±0.10
	3	66.84±0.05	70.45±0.11	73.42±0.11	77.51±0.11	83.71±0.05
	4	74.96±0.02	78.82±0.03	81.88±0.05	86.28±0.03	90.04±0.04
	5	82.66±0.11	86.50±0.03	89.19±0.03	92.73±0.03	95.85 ±0.11
	6	89.96±0.10	92.27±0.10	94.2±0.10	97.30±0.10	**99.4±0.10***
Lipase	**Time(h) ↓**	**45**	**50**	**55**	**60**	**65**
RMIM	0	0	0	0	0	0
	1	36.50±0.02	42.73±0.04	47.47±0.03	50.73±0.03	54.27±0.02
	2	57.67±0.10	61.50±0.06	64.28±0.05	68.04±0.03	71.14±0.10
	3	65.74±0.05	68.91±0.11	73.42±0.11	75.98±0.11	81.41±0.05
	4	73.81±0.03	76.9±0.02	81.12±0.02	84.74±0.01	87.35±0.02
	5	81.12±0.11	84.58±0.03	87.65±0.03	91.57±0.03	93.88±0.11
	6	86.50±0.10	89.96±0.10	92.66±0.10	95.76±0.10	97.47±0.10
Lipase	**Time(h) ↓**	**45**	**50**	**55**	**60**	**65**
TLIM	0	0	0	0	0	0
	1	34.58±0.02	38.88±0.03	43.63±0.03	46.88±0.03	50.42±0.02
	2	53.82±0.1	57.65±0.05	60.43±0.06	64.19±0.05	67.30±0.1
	3	61.90±0.05	65.07±0.11	69.58±0.11	72.13±0.11	77.56±0.05
	4	69.96±0.05	73.05±0.02	77.27±0.03	80.90±0.04	84.65±0.03
	5	77.28±0.11	80.73±0.03	83.81±0.03	87.73±0.03	90.04±0.11
	6	82.65±0.10	86.11±0.10	88.81±0.10	91.92±0.10	95.17±0.10

*Values are represented as mean ± SEM. n=3 ($p<0.05$)

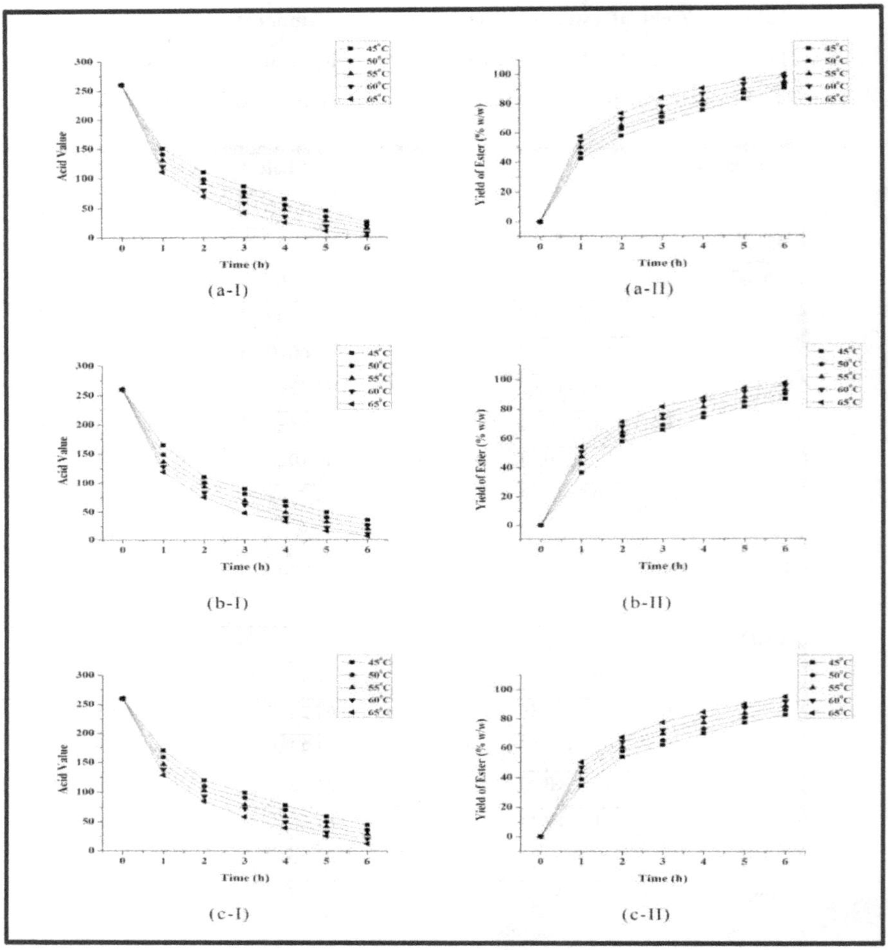

Figure 5.(a-I, b-I and c-I) Changes in Acid value at different reaction temperature (45^0C to 65^0C) at a constant molar ratio (1:1) of capric acid and cetyl alcohol using NS 435,RMIM and TLIM as biocatalyst (10%w/w) respectively at different time intervals (0-6 h); and (a-II,b-II and c-II) Yield of ester at different reaction temperature (45^0C to 65^0C) using NS 435,RMIM and TLIM as biocatalyst (10% w/w) respectively at a constant molar ratio (1:1) of capric acid and cetyl alcohol at a different reaction time (0-6h). Values are represented as mean ± SEM.n=3 ($p<0.05$)

Table 12.Changes in Acid value at different reaction molar ratio (1:1,1:2 and 1:3) of capric acid and cetyl alcohol using NS 435,RMIM and TLIM as biocatalyst at a constant lipase loading (10%w/w) and at a constant reaction temperature (65^0C) at different time intervals (0 -6 h)

Lipase	Time(h) ↓	Acid Value		
		Molar Ratio (Capric acid and Cetyl alcohol)		
		1:1	**1:2**	**1:3**
NS435	0	260.56±0.20	260.06±0.10	260.15±0.10
	1	110.89±0.02	130.07±0.15	149.06±0.15
	2	70.03±0.10	85.09±0.05	95.03±0.05
	3	42.34±0.05	56.89±0.10	65.08±0.10
	4	25.89±0.02	35.09±0.12	46.08±0.12
	5	10.89±0.12	20.09±0.05	32.08±0.05
	6	**1.56±0.10***	9.50±0.06	19.5±0.06
Lipase	**Time(h) ↓**	**1:1**	**1:2**	**1:3**
RMIM	0	260.56±0.20	260.06±0.10	260.15±0.10
	1	118.89±0.02	139.07±0.15	156.06±0.15
	2	75.03±0.10	90.09±0.05	105.03±0.05
	3	48.34±0.05	58.89±0.10	69.08±0.10
	4	32.89±0.03	43.09±0.12	52.08±0.12
	5	15.89±0.11	25.09±0.05	38.08±0.05
	6	6.56±0.10	15.50±0.06	25.50±0.06
Lipase	**Time(h) ↓**	**1:1**	**1:2**	**1:3**
TLIM	0	260.56±0.2	260.06±0.10	260.15±0.10
	1	128.89±0.02	149.07±0.15	166.06±0.15
	2	85.03±0.10	100.10±0.05	115.03±0.05
	3	58.34±0.05	68.89±0.10	79.08±0.10
	4	39.89±0.03	53.09±0.12	62.08±0.12
	5	25.89±0.11	35.09±0.05	48.08±0.05
	6	12.56±0.10	25.50±0.06	35.50±0.06

*Values are represented as mean ± SEM. n=3 ($p<0.05$)

Table 13.Yield of ester at different molar ratio (1:1,1:2 and 1:3) of capric acid and cetyl alcohol using NS 435,RMIM and TLIM as biocatalyst in a constant lipase loading (10% w/w) at a constant reaction temperature (65^0C) at different reaction time intervals (0-6h)

Lipase	Time(h) ↓	Yield of Ester(%w/w)		
		Molar Ratio (Capric acid and Cetyl alcohol)		
		1:1	1:2	1:3
NS435	0	0	0	0
	1	57.35±0.02	49.97±0.15	42.67±0.15
	2	73.06±0.10	67.27±0.05	63.45±0.05
	3	83.71±0.05	78.12±0.11	74.97±0.10
	4	90.04±0.03	86.50±0.12	82.28±0.12
	5	95.81±0.11	92.27±0.05	87.66±0.05
	6	**99.4±0.10***	96.34±0.06	92.50±0.06
Lipase	**Time(h) ↓**	**1:1**	**1:2**	**1:3**
RMIM	0	0	0	0
	1	54.27±0.02	46.51±0.15	39.97±0.15
	2	71.14±0.10	65.35±0.05	59.60±0.05
	3	81.40±0.05	77.35±0.10	73.43±0.10
	4	87.35±0.03	83.42±0.12	79.96±0.12
	5	93.88±0.11	90.35±0.05	85.35±0.05
	6	97.47±0.10	94.03±0.06	90.19±0.06
Lipase	**Time(h) ↓**	**1:1**	**1:2**	**1:3**
TLIM	0	0	0	0
	1	50.42±0.02	42.66±0.15	36.13±0.15
	2	67.29±0.05	61.50±0.10	55.75±0.10
	3	77.56±0.10	73.50±0.05	69.58±0.05
	4	84.65±0.05	79.58±0.10	76.12±0.10
	5	90.04±0.11	86.50±0.05	81.50±0.05
	6	95.16±0.10	90.19±0.06	86.34±0.06

*Values are represented as mean ± SEM. n=3 ($p<0.05$)

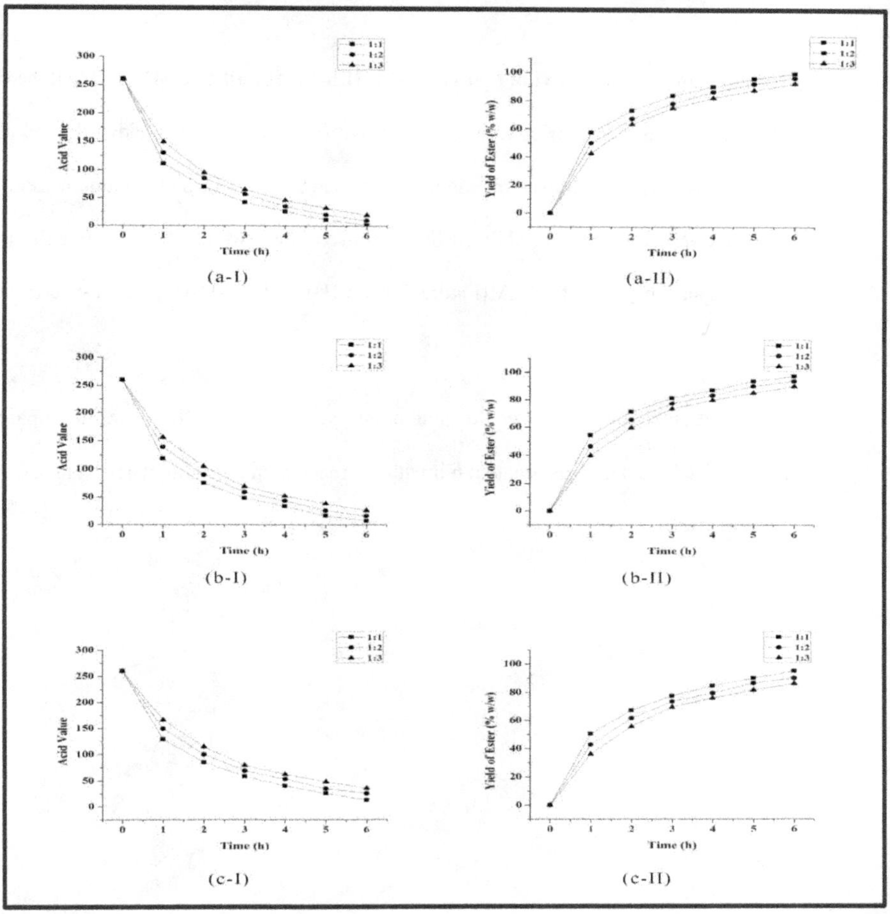

Figure 6. (a-I, b-I and c-I) Changes in Acid value at different reaction molar ratio (1:1,1:2 and 1:3) of capric acid and cetyl alcohol using NS 435,RMIM and TLIM as biocatalyst at a constant lipase loading (10%w/w) and at a constant reaction temperature (65^{0}C) at different time intervals (0-6 h); and (a-II,b-II and c-II) Yield of ester at different molar ratio (1:1,1:2 and 1:3) of capric acid and cetyl alcohol using NS 435,RMIM and TLIM as biocatalyst in a constant lipase loading (10% w/w) at a constant reaction temperature (65^{0}C) at different reaction time (0-6h). Values are represented as mean ± SEM. n=3 ($p<0.05$)

Table 14 and Table 15 (vide Figure 7 a-I,a-II,b-I,b-II and c-I, c-II),depicted the effect of different lipases and their concentrations on the reaction of capric acid and cetyl alcohol . Figure 7 a-I,a-II, also showed the highest decrease in acid value and maximum yield of ester at 10% (w/w) NS 435 immobilized lipase concentration which was significantly ($p<0.05$) higher than the yield of RMIM(Figure 7 b-I,b-II) and TLIM (Figure 7c-I, c-II)catalyzed reaction.

Therefore, reaction parameter for maximum yield of CCAPR ester, the lipase being immobilized NS 435, was found at 6 h with the reactant molar ratio of 1:1 and 65^{0}C reaction temperature.

Table 14 .Changes in acid value at different lipase loading (4,6,8,10 and 12% w/w) using NS 435,RMIM and TLIM as biocatalyst at a constant molar ratio (1:1) of capric acid and cetyl alcohol at constant temperature (65 ^{0}C) with different time intervals (0 h to 6 h)

Lipase	Time(h) ↓	Acid Value				
		Lipase loading (%w/w)				
		4	6	8	10	12
NS435	0	260.05±0.10	262.09±0.10	260.5±0.10	260.56±0.10	260.09±0.10
	1	150.09±0.03	138.05±0.02	129.06±0.15	110.89±0.15	119.09±0.03
	2	100.05±0.05	92.76±0.05	85.06±0.2	70.03±0.2	75.08±0.05
	3	75.06±0.11	64.89±0.12	58.08±0.05	42.34±0.05	50.01±0.11
	4	55.67±0.03	45.08±0.15	38.45±0.03	25.89±0.03	33.01±0.03
	5	40.09±0.03	29.09±0.15	22.07±0.15	10.89±0.15	17.04±0.03
	6	25.08±0.10	15.06±0.12	8.56±0.10	**1.56±0.10***	6.02±0.10
Lipase	**Time(h) ↓**	**4**	**6**	**8**	**10**	**12**
RMIM	0	260.05±0.10	262.10±0.10	260.5±0.10	260.56±0.10	260.09±0.10
	1	165.09±0.03	155.05±0.03	145.06±0.15	125.19±0.15	135.09±0.03
	2	120.05±0.05	110.76±0.05	100.06±0.20	80.03±0.20	90.08±0.04
	3	85.06±0.11	75.09±0.12	65.08±0.05	45.34±0.05	55.01±0.11
	4	65.67±0.03	55.08±0.15	45.05±0.02	28.89±0.03	38.01±0.05
	5	50.09±0.03	40.09±0.15	30.07±0.15	15.09±0.15	22.04±0.03
	6	35.08±0.10	25.06±0.12	18.56±0.10	6.56±0.10	12.02±0.10
Lipase	**Time(h) ↓**	**4**	**6**	**8**	**10**	**12**
TLIM	0	260.05±0.10	262.09±0.10	260.5±0.10	260.56±0.10	260.09±0.10
	1	168.09±0.03	158.05±0.02	148.06±0.15	128.89±0.15	138.09±0.04
	2	125.05±0.06	115.76±0.05	107.06±0.2	85.03±0.2	95.08±0.05
	3	90.06±0.11	83.09±0.12	75.08±0.05	58.34±0.05	65.01±0.11
	4	65.67±0.02	58.08±0.15	50.05±0.03	39.89±0.04	45.01±0.03
	5	50.09±0.03	45.09±0.15	39.07±0.15	25.89±0.15	32.04±0.03
	6	40.08±0.10	33.06±0.10	28.56±0.10	12.56±0.10	20.02±0.10

*Values are represented as mean ± SEM. n=3 ($p<0.05$)

Table 15.Yield of ester at different catalyst loading (4,6,8,10 and 12% w/w) using NS 435,RMIM and TLIM as biocatalyst respectively at a constant molar ratio (1:1) of capric acid and cetyl alcohol and at constant reaction temperature (65^{0}C) with different reaction time (0-6h) intervals

Lipase	Time(h) ↓	Yield of Ester (%w/w) Lipase loading(%w/w)				
		4	6	8	10	12
NS435	0	0	0	0	0	0
	1	42.27±0.03	46.90±0.03	50.36±0.15	57.35±0.02	54.19±0.02
	2	61.52±0.05	64.32±0.05	67.28±0.20	73.06±0.10	71.12±0.10
	3	71.13±0.11	75.04±0.12	77.66±0.05	83.71±0.05	80.76±0.05
	4	78.58±0.02	82.66±0.15	85.21±0.03	90.04±0.02	87.30±0.03
	5	84.58±0.03	88.81±0.15	91.51±0.15	95.81±0.11	93.44±0.11
	6	90.35±0.10	94.20±0.12	96.70±0.10	**99.4±0.10***	97.68±0.10
Lipase	**Time(h) ↓**	**4**	**6**	**8**	**10**	**12**
RMIM	0	0	0	0	0	0
	1	36.50±0.04	40.36±0.03	44.21±0.15	51.85±0.02	48.04±0.03
	2	53.83±0.05	57.40±0.05	61.51±0.2	69.22±0.1	65.35±0.05
	3	67.28±0.11	71.12±0.12	74.97±0.05	82.56±0.05	78.84±0.11
	4	74.74±0.025	78.82±0.15	82.67±0.03	88.88±0.04	85.38±0.03
	5	80.73±0.03	84.60±0.15	88.43±0.15	94.20±0.11	91.52±0.03
	6	86.50±0.10	90.36±0.11	92.86±0.12	97.47±0.10	95.37±0.09
Lipase	**Time(h) ↓**	**4**	**6**	**8**	**10**	**12**
TLIM	0	0	0	0	0	0
	1	35.35±0.04	39.21±0.03	43.05±0.15	50.43±0.02	46.88±0.03
	2	51.90±0.05	55.47±0.06	58.82±0.2	67.30±0.1	63.43±0.05
	3	65.36±0.11	68.04±0.12	71.12±0.05	77.56±0.05	75.01±0.11
	4	74.74±0.03	77.66±0.15	80.75±0.02	84.65±0.04	82.69±0.03
	5	80.73±0.03	82.65±0.15	84.97±0.15	90.04±0.11	87.67±0.03
	6	84.58±0.10	87.28±0.12	89.01±0.10	95.16±0.11	92.30±0.12

*Values are represented as mean ± SEM. n=3 ($p<0.05$)

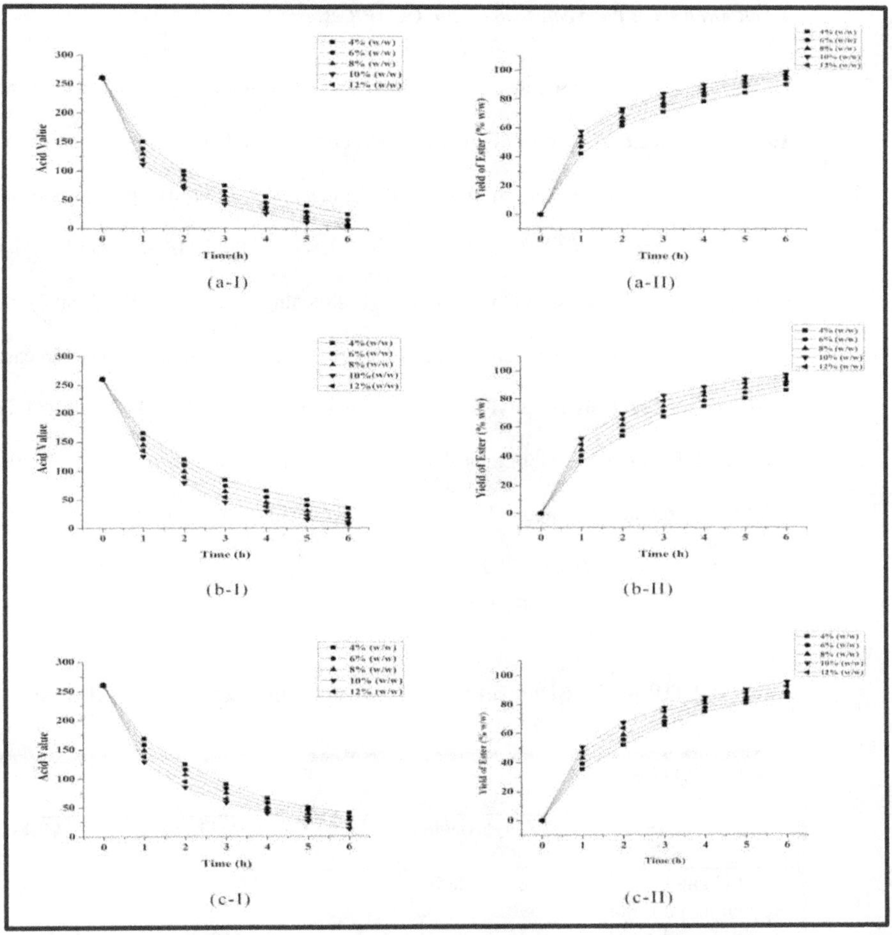

Figure 7. (a-I, b-I and c-I) Changes in acid value at different lipase loading (4,6,8,10 and 12% w/w) using NS 435,RMIM and TLIM as biocatalyst at a constant molar ratio (1:1) of capric acid and cetyl alcohol at constant temperature (65 ^{0}C) with different time intervals (0-6 h); and (a-II ,b-II and c-II) Yield of ester at different lipase loading (4,6,8,10 and 12% w/w) using NS 435,RMIM and TLIM as biocatalyst respectively at a constant molar ratio (1:1) of capric acid and cetyl alcohol and at constant reaction temperature (65^{0}C) with different reaction time (0-6h) intervals. Values are represented as mean ± SEM. n=3 ($p<0.05$)

FTIR analysis of pure capric acid and CCAPR ester:

FTIR spectra of pure capric acid ,cetyl alcohol and CCAPR ester was presented in the **Table 16 (vide Figure 8 a, b and c).**Figure 8 (a) described the *FTIR* spectra of capric acid. O–H stretching vibration (2922.8 cm^{-1}), C=O stretching vibration (1701 cm^{-1}) was observed in case of *FTIR* spectra of capric acid. Figure 8(b) described the FTIR spectra of cetyl alcohol. O–H stretching vibration (3325.1 cm^{-1}),C–H stretching vibration (2918.2 cm^{-1}) and C–O stretching (1463 cm^{-1},1064.3 cm^{-1}, and 719.7 cm^{-1}) was found to be observed in case of the *FTIR* spectral analysis of cetyl alcohol. FTIR spectra of the CCAPR ester (Figure 8 c) showed C–H stretching vibrations (2920.0 cm^{-1}),O–H stretching vibrations (2850.2cm^{-1}), C=O ester stretching vibration (1732 cm^{-1}),C–O stretching vibrations (1326.2cm^{-1} to 1473.1 cm^{-1}) and C–H bending (720.0 $cm^{-1}$to 1201 cm^{-1}) which confirmed the CCAPR ester formation.

Table 16.FTIR peak assignment for cetyl alcohol, capric acid and CCAPR ester

Group of class	**Range(cm^{-1})**		
	Cetyl Alcohol	**Capric Acid**	**CCAPR ester**
O-H and C-H stretching vibration	2910-2850	2922.8	2920.0-2850
O-H bending	1100	945-1418	943-1415
C-O stretching vibrations	1463- 719.7	1467-1353	1473.1 -1326.2
C=O stretching vibration	-	1701	1732
C-H rocking vibration	1470-717	-	1468-715
C-H bending	-	1218-456	1201-720

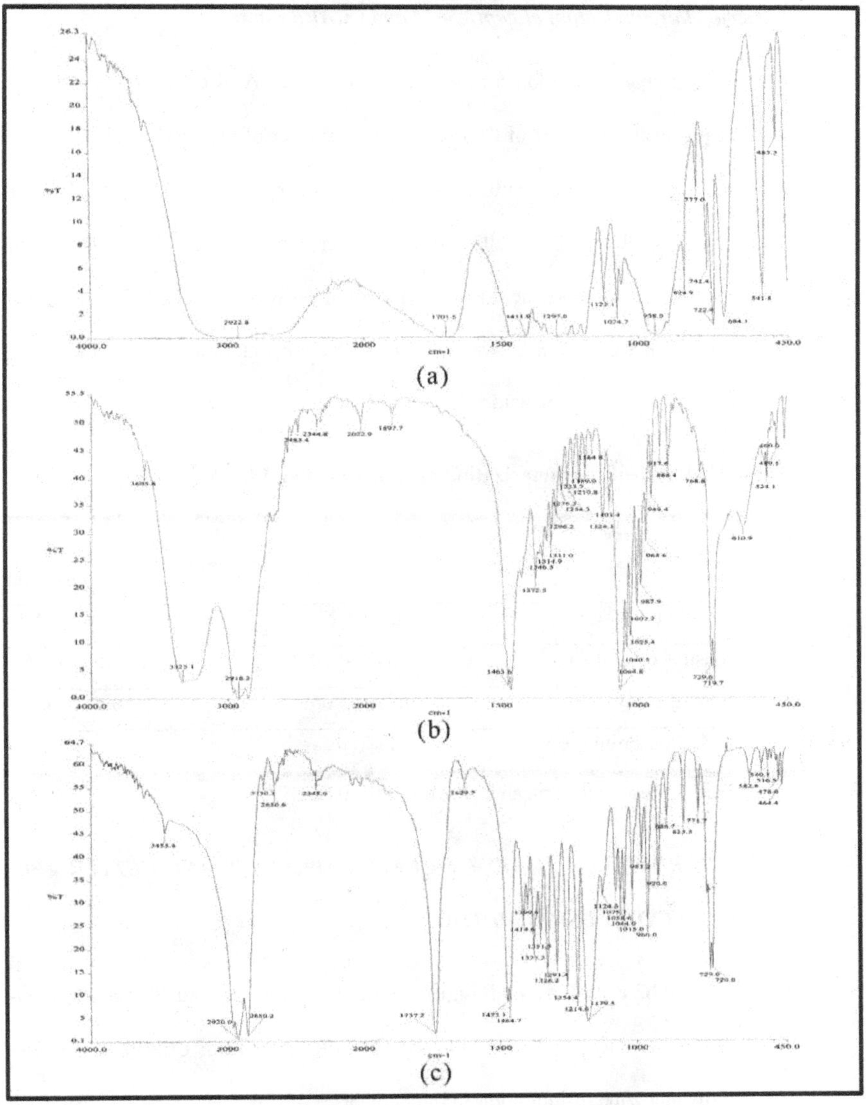

Figure 8. FTIR spectra of (a) Capric acid,(b) Cetyl alcohol and (c) CCAPR ester

Physical characterization of capric acid and CCAPR ester:

The physical characterizations of the capric acid and CCAPR ester were listed in the **Table 17**. The physical properties of the capric acid and ester of the cetyl alcohol and capric acid obtained were comparable with the results obtained by other researchers. The flash point was obtained above the 280^{0}C. The viscosity of capric acid is 4.372cP at 40^{0}C temperature, whereas for ester of capric acid and cetyl alcohol , it was 4.02 at 40 °C. The density of the capric acid and CCAPR ester was almost similar. Slip melting point of the CCAPR ester was quite lower than capric acid.

Table 17 .Physical characterization of capric acid and CCAPR ester

Properties	Values	
	Capric acid	**CCAPR ester**
Flash Point (^{0}C)	130±0.47	Above 280
Viscosity(cP at 40^{0}C)	4.372±0.10	4.02±0.15
Density (g/cm^3) at 30^{0}C	0.894±0.05	0.851±0.03
Slip Melting Point (^{0}C)	23±0.02	21.5±0.05

Values are represented as mean ± SEM. n=3(p<0.05)

ANTI-MICROBIAL ACTIVITY OF CETYL CAPRYLATE AND CETYL CAPRATE BY AGAR WELL DIFFUSION METHOD

Figure 9 (a, b, c and d) and Figure10 (a, b, c and d) and Table 18 described the antimicrobial activity of the CCAPL and CCAPR ester. Both CCAPL and CCAPR ester showed distinct zone of inhibition in the both cases of *E.coli* and *S.aureus*. But CCAPL and CCAPR both the esters exhibited better antimicrobial activity in case of *S.aureus* in comparison to *E.coli*. These esters might be potential to the *S.aureus*.

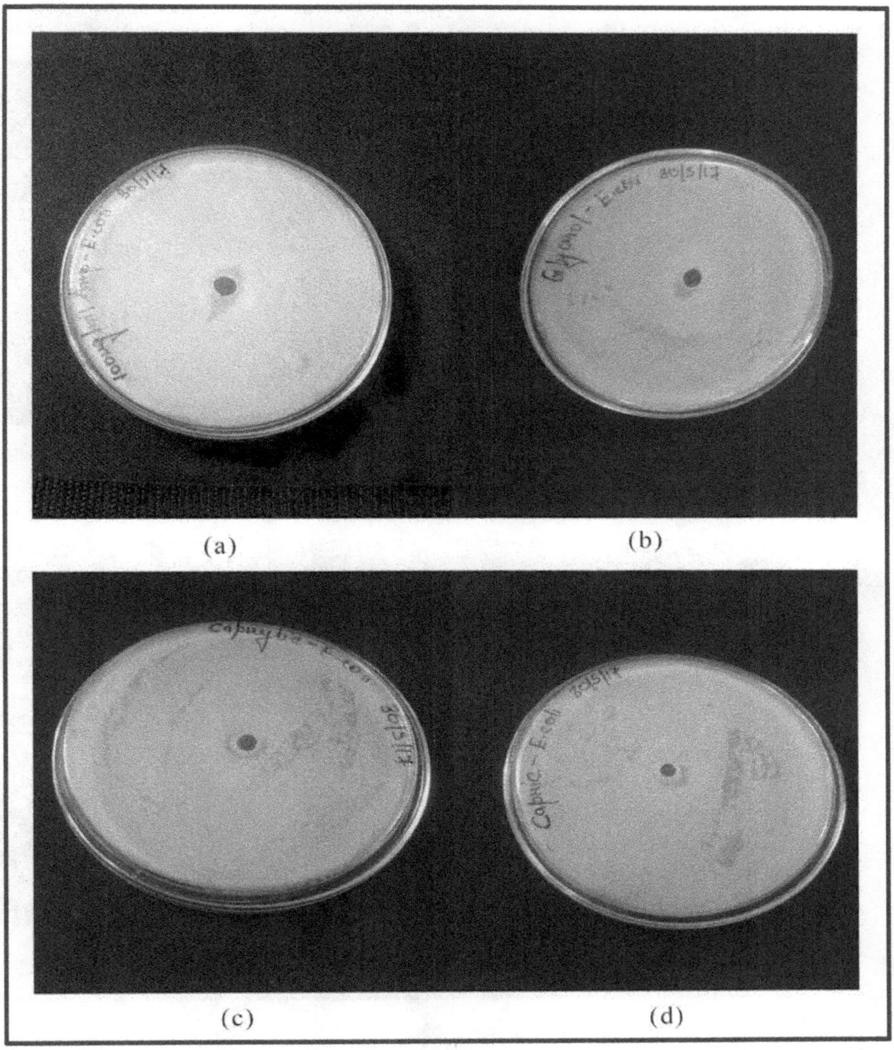

Figure 9. Representative picture of inhibitory zones of *E.coli* against different edible plasticizers incorporated into the well diffusion plates. (a) Control (b) Glycerol (c) CCAPL (d) CCAPR

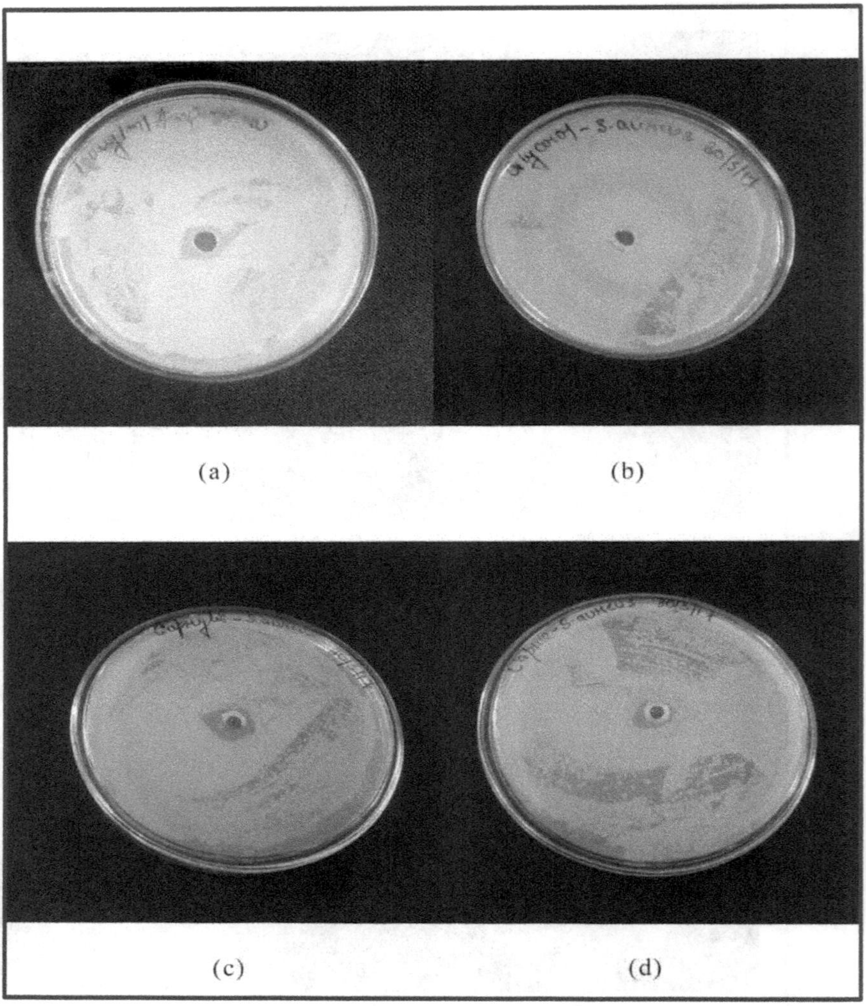

Figure 10.Representative picture of inhibitory zones of *S.aureus* against different edible plasticizers incorporated into the well diffusion plates. (a) Control (b) Glycerol (c) CCAPL (d) CCAPR

Table 18.Antimicrobial activity of CCAPL and CCAPR against the test microorganisms

Strains	Samples	ZoI(mm)
E.coli	Ampicillin	1.56 ± 0.5
S.aureus	Ampicillin	1.70±0.02[a]
E.coli	CCAPL*	1.36±0.02
S. aureus	CCAPL	1.96±0.09[b]
E.coli	CCAPR#	1.23±0.06
S. aureus	CCAPR	1.63±0.25[c]

*Values are represented as mean ± SEM. n=3. Superscripts (a to c) indicate significant differences in results ($p < 0.05$) between ZOI of esters [CCAPL and CCAPR] against *E.coli* and *S.aureus*.

#CCAPR- Cetyl caprate, #CCAPL-Cetyl caprylate

PERFORMANCE EVALUATION OF THE CETYL CAPRYLATE AND CETYL CAPRATE ESTER PLASTICIZED CARBOXYMETHYL CELLULOSE (CMC) FILMS

X-ray diffraction analysis of the CMC films:

Figure 11(a-I to a-V) and (b-I to b-V) explained the XRD patterns of pure CMC films, CMC films containing CCAPL and CCAPR as plasticizers respectively. Glycerol was used as conventional edible plasticizer. Figure 11(a-I and b-I) showed the XRD pattern of pure CMC film which was amorphous in nature but in case of the CMC films containing glycerol as conventional glycerol as plasticizer (Figure 11 a-II) showed a bell shaped curve at 20° of 2θ position. Same patterns of curves were observed in Figure 11 (a-III to a-V) when CCAPL was incorporated in the CMC films at different concentration by replacing glycerol.

Figure 11 (b-I to b-V) depicted the XRD patterns of Pure CMC film, CMC films containing glycerol as conventional plasticizer and CMC films containing CCAPR ester at different concentration. This Figure 11 (b-III to b-V) also showed the same XRD pattern as the glycerol plasticized CMC films. Hence, CCAPL and CCAPR possessed the same performance as glycerol.

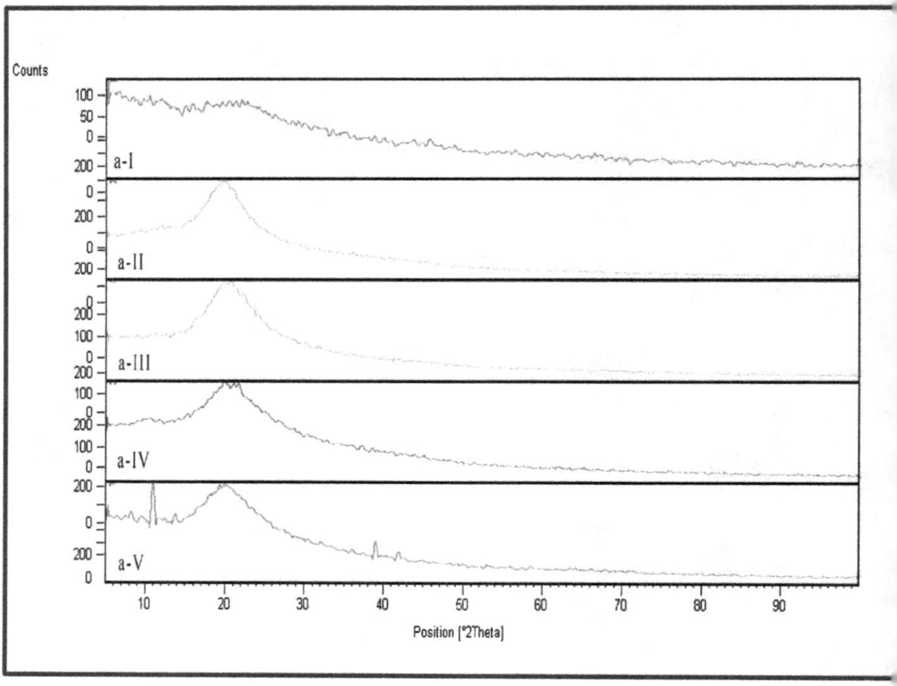

Figure 11. (a-I to a-V) X-ray diffraction analysis of Pure CMC film, CMC film containing glycerol and CCAPL ester at different concentration

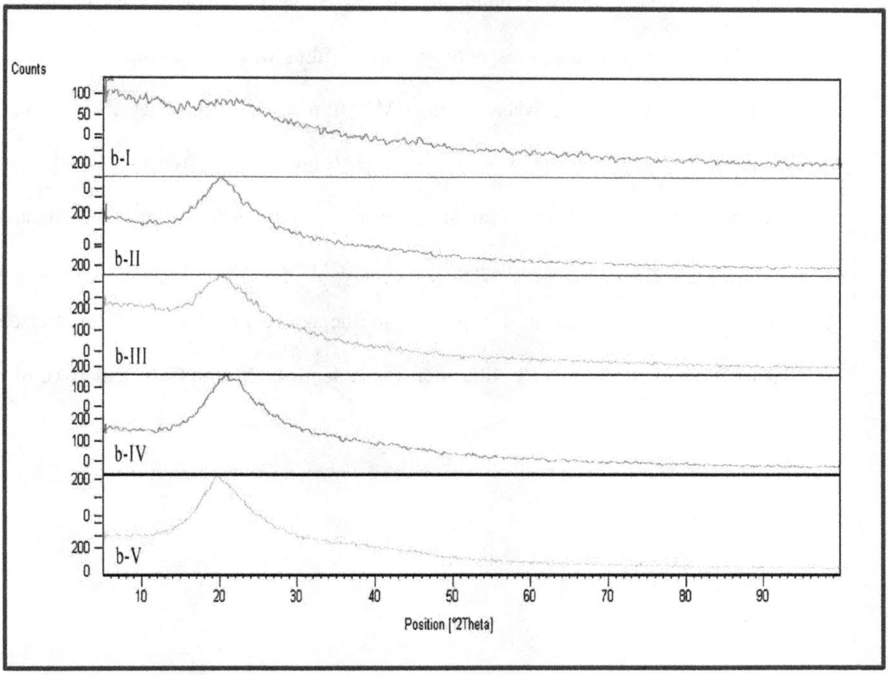

Figure 11. (b-I to b-V). X-ray diffraction analysis of Pure CMC film, CMC film containing glycerol and CCAPR ester at different concentration

Scanning Electron Microscopy:

Figure 12 (a, b, c, d, e) and Figure 13 (a, b, c, d, e) revealed the SEM images of the pure CMC film, CMC films containing glycerol as conventional plasticizer, CMC films with CCAPL and CCAPR esters respectively . CMC films without plasticizers showed (Figure 12 a) a smooth surface whereas the CMC film with glycerol as conventional edible plasticizers showed a quite rough surface (Figure 12 b).When CCAPL ester was incorporated into the CMC film at different concentration (Figure 12 c, d, and e) by replacing the glycerol, the surface of the films (CMC-CCAPL 1,CMC-CCAPL 2 and CMC-CCAPL3)were smoother in comparison to the conventional glycerol plasticized films. Hence it could be said that CCAPL ester was quite miscible with CMC and glycerol.

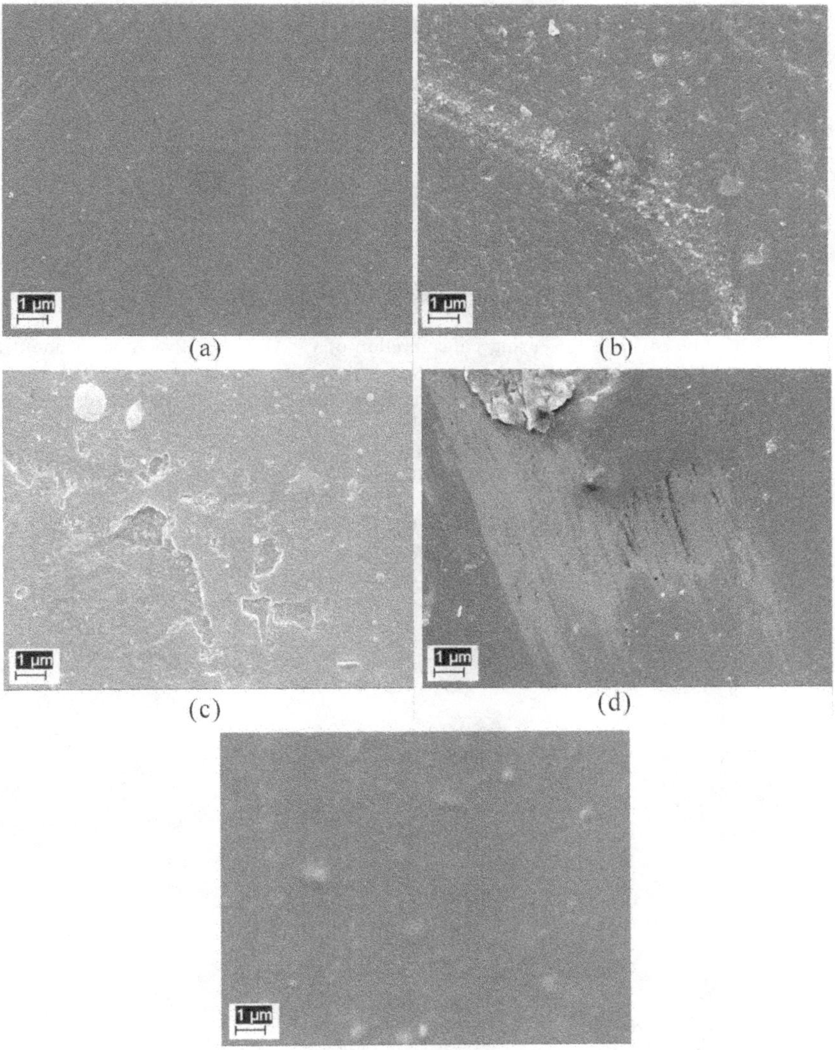

(a) (b) (c) (d) (e)

Figure 12. SEM images of CMC film surfaces {(a) Pure CMC film,(b) CMC film with Glycerol, (c to e) CMC with CCAPL ester [at different concentrations] as plasticizer}

SEM images of the CMC films containing CCAPR as plasticizing agent was presented in the **Figure 13 (a, b, c, d, and e).** CMC films without plasticizers showed (Figure 13 a) a smooth surface whereas the CMC film with glycerol as conventional edible plasticizers showed a quite rough surface (Figure 13 b). CMC film containing glycerol as a plasticizer in a large proportion and CCAPR at a minor proportion showed (Figure 13 c) quite rough surface in comparison to the pure CMC film (Figure 13 a).But in case of Figure 13(d) , a different kind of surface was observed which may be caused due to the interaction amongst CMC, glycerol and increasing concentration of CCAPR. Finally a quite smoother surface was found in case of Figure 13 (e).This phenomenon might be caused due to the miscibility of capric acid at a higher concentration with CMC and with the least proportion of glycerol.

Therefore the SEM images (Figure 12 and Figure 13) of CMC films containing CCAPL and CCAPR as a plasticizer could be good replacement.

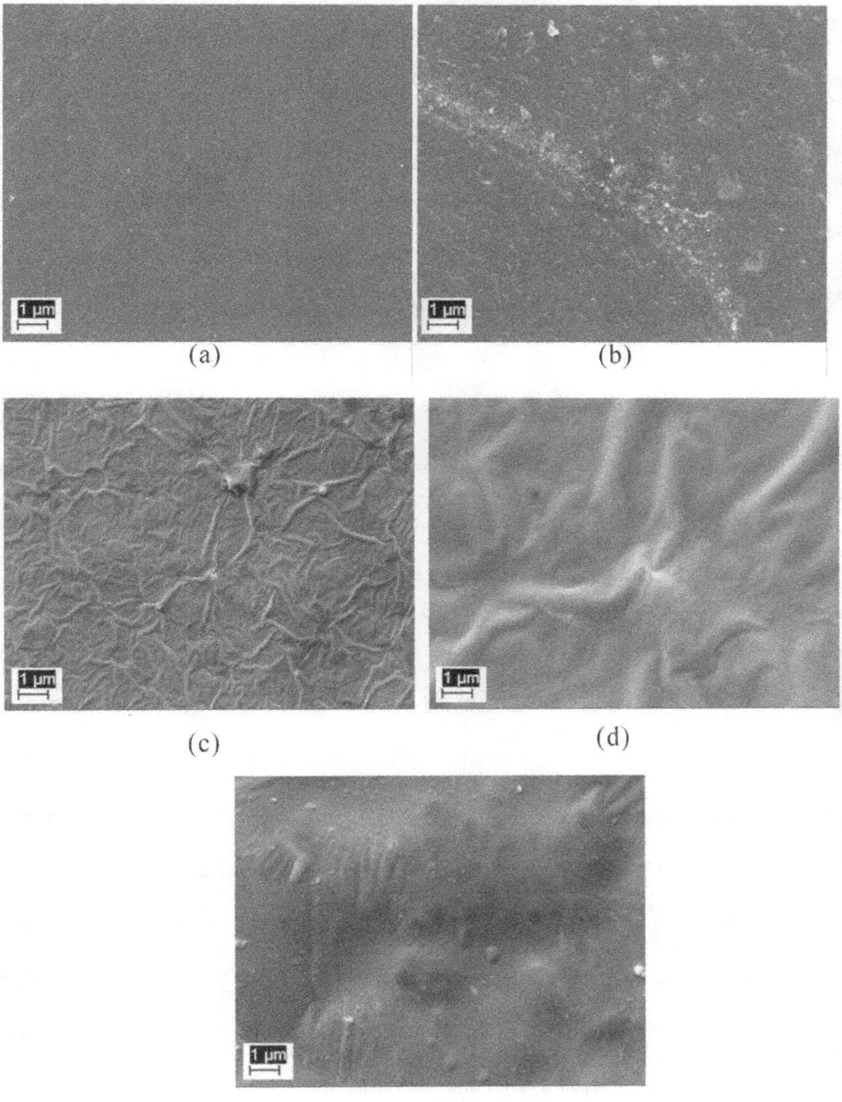

Figure 13. SEM images of CMC film surfaces {(a) Pure CMC film,(b) CMC film with Glycerol, (c to e) CMC with CCAPR ester [at different concentrations] as plasticizer }

Mechanical performance of CMC films:

Results of the tensile tests of the different CMC films are shown in **Table 19**. The high tensile strength and Young's modulus values and low elongation value of the control sample could be attributed to hydrogen bonds between the CMC chains. These bonds contribute to the high cohesiveness and low flexibility of the unplasticized films. Glycerol ,CCAPL ester and CCAPR ester containing CMC films showed an increase in elongation property, decrease in tensile strength and Young's modulus with increasing concentration of glycerol or CCAPL or CCAPR .However, the effect of CCAPL and CCAPR on the mechanical properties of the CMC films was stronger as shown in Table 19, the tensile strength of pure CMC film was 22.01 MPa for the control (unplasticized) sample and decreases significantly ($p< 0.05$) to 2.09Mpa and 5.17MPa for the CMC-CCAPL 3 film and CMC-CCAPR 3 films respectively. But the films containing only glycerol showed (Table 19) poor tensile strength in comparison to CCAPL and CCAPR.

Similarly, the Young's modulus of the films decreases from 580 MPa for the control sample, whereas CMC-CCAPL 3 showed 70.02MPa decreasing effect of CCAPL on the Young's modulus value was lower than glycerol and that value was 90.41MPa for the CMC films containing CCAPR i.e. CMC-CCAPR 3.Hence lipid esters addition induced the development of a heterogeneous film structure featuring discontinuities in the polymer network. The result is a decrease in tensile strength and Young's modulus values and an increase of film flexibility. In the presence of CCAPL or CCAPR, a small increase in the glycerol level results in a large drop in tensile strength and Young's modulus, a significant increase in elongation properties.

Table 19.Mechanical properties of CMC films with or without plasticizer

Sample	Young's Modulus (MPa)	Tensile Strength (MPa)	Elongation at break (%)
CMC (control)	580±0.52	22.01±0.15	6.70±0.05
CMC-GLY[a]	72.76±1.09	13.12±0.24	27.56±0.14
CMC-CCAPL1[b]	89.02±1.59	9.86±0.14	30.67±0.13
CMC-CCAPL2	90.07±0.99	4.37±0.31	35.09±0.28
CMC-CCAPL3	70.08±0.93	2.09±0.21	45.09±0.08
CMC-CCAPR 1[c]	397.03±2.09	9.65±0.23	24.09±0.10
CMC -CCAPR2	79.53±3.23	7.73±0.32	29.80±0.07
CMC-CCAPR3	90.41±1.19	5.17±0.19	32.09±0.37

Values are represented as Mean ± SEM. n=3($p < 0.05$)

[a]CMC-GLY- Carboxymethyl cellulose + glycerol

[b] CMC-CCAPL- Carboxymethyl cellulose + cetyl caprylate ester

[c] CMC-CCAPR-Carboxymethyl cellulose + cetyl caprate ester

Thermal properties:

Thermogravimetric (TG) and Differential Thermal Analysis (DTA) curves of the pure CMC films and plasticized (glycerol, CCAPL and CCAPR) films were presented in **Figure 14 (a and b)** respectively. Pure CMC film and CMC films containing glycerol (Figure 14 a), showed the same thermal stability. Same trend was also maintained in the case of the CMC films containing CCAPL and CCAPR as non-conventional edible plasticizer. Same kind of glass transition temperature was also observed in all cases of plasticized and non plasticized CMC films (Figure 14 b).

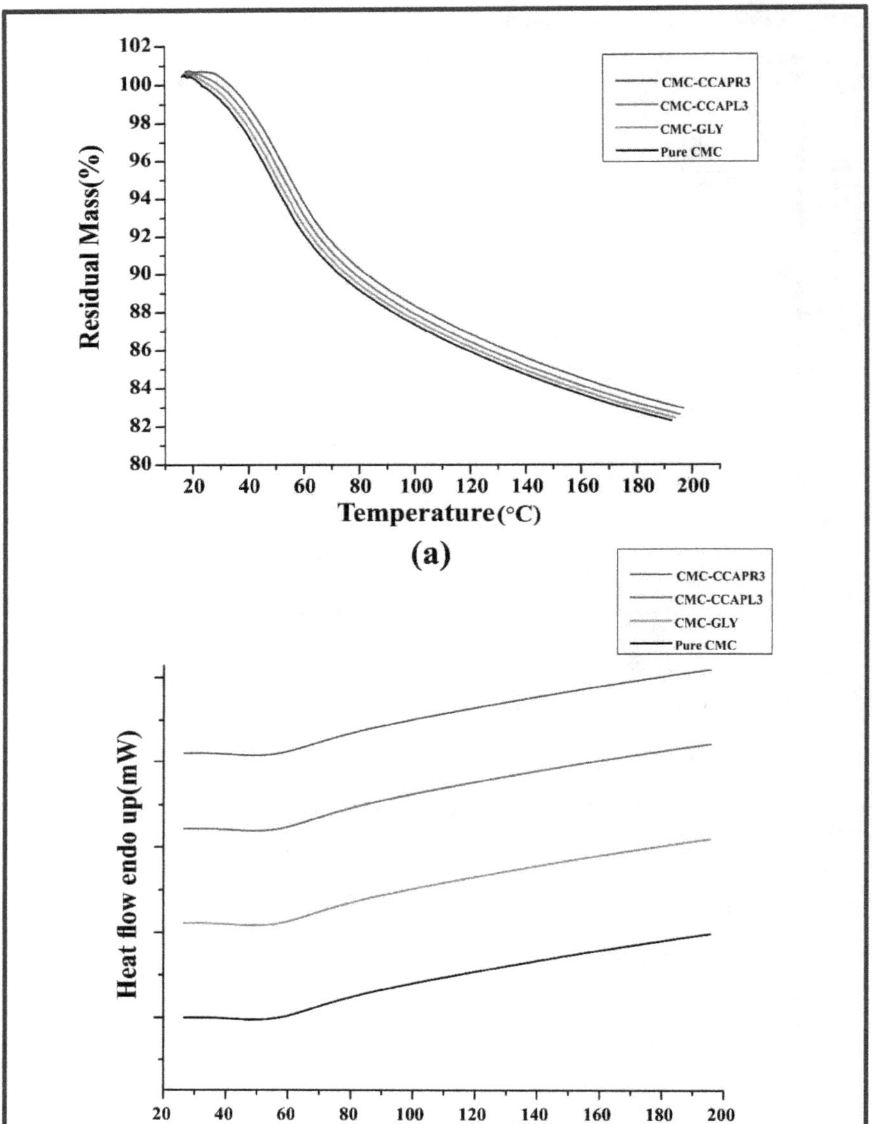

Figure 14. TG DTA curves of a) Thermogravimetric analysis & (b) Differential thermal analysis , pure CMC film , CMC-Glycerol , CMC-CCAPL 3 , CMC-CCAPR 3

Water Vapor permeability (WVP):

Values of WVP including pure CMC films and the films with varying amount of glycerol, CCAPL and CCAPR contents were presented in **Table 20**. WVP of the pure CMC films was affected by glycerol, CCAPL and CCAPR esters. The pure CMC films showed lowest value of WVP while the value increased by the addition of glycerol. Commonly, transmission of water vapor through a hydrophilic film depends upon both the diffusivity and solubility of water molecules in the film matrix. Insertion of glycerol molecules between the polymer chains enhance the water vapor diffusivity through the film and accelerated the water vapor transmission. Hydrophilic glycerol molecules induce the adsorption of water molecules and could contribute to the increase of WVP of the CMC film. But each CMC films containing CCAPL and CCAPR as plasticizer showed lowest water vapor permeable property and highest barrier properties at their highest concentration into the CMC films. This phenomenon could be attributed to the highly hydrophobic nature of CCAPL and CCAPR esters present in the film composition. Moreover, the polymeric chains in the film forming solution could partially be immobilized at the edge with the emulsified lipid particles. Subsequently, the polymeric chains became immobile by reducing the diffusibility of water through the CMC chains interface and leading to a decrease in WVP [1,2].

Table 20.Water Vapor Permeability properties of CMC film with or without plasticizer

Sample	Water Vapor permeability ($10^{-5}g/cm^2/24$ h)
Pure CMC	0.85±0.02
CMC-GLY[a]	1.85±0.03
CMC-CCAPL 1[b]	1.25±0.01
CMC-CCAPL 2	1.02±0.05
CMC -CCAPL 3	**0.42±0.02***[*]
CMC-CCAPR 1[c]	1.39±0.04
CMC- CCAPR 2	1.34±0.02
CMC-CCAPR 3	**0.72±0.04**[#]

Values are represented as Mean ± SEM. n=3.*# Superscripts in the same column indicate significant differences in results ($p < 0.05$)

[a]CMC-GLY- Carboxymethyl cellulose + glycerol

[b] CMC-CCAPL- Carboxymethyl cellulose + cetyl caprylate ester

[c] CMC-CCAPR-Carboxymethyl cellulose + cetyl caprate ester

Moisture uptake of CMC films:

Table 21 described the moisture uptake during conditioning at 97%RH for 48 h. The unplasticized pure CMC films showed lower moisture absorption than the glycerol-plasticized CMC films. Though cellulose derivatives are typical hydrophilic polymers and its medium crystallinity reduces the water absorbency of edible CMC films. Nevertheless, mishmash of glycerol into the CMC matrix was able to increase the moisture uptake capacity of CMC films. Hydrophilic nature of glycerol promotes its immobilization between CMC chains and leads to the interaction with water molecules. CMC films containing

CCAPL and CCAPR esters respectively showed (Table 21) reduced moisture uptake with their use in increased concentration into the CMC films. In addition, in the presence of glycerol, CCAPL/CCAPR was able to reduce the moisture uptake of the CMC films significantly ($p < 0.05$). Hydrophobic nature of CCAPL and CCAPR esters causes the reduction of the moisture uptake capacity of CMC macromolecules. Furthermore, addition of wax esters probably causes the immobilization of biopolymer chains which in turn reduces their tendency for interaction with water molecules [1].

Table 21.Moisture uptake of CMC films with or without plasticizers

Samples	Moisture Absorption(%w/w)
Pure CMC	40.09±0.1
CMC-GLY[a]	140.01±0.2
CMC -CCAPL1[b]	100.50±0.15
CMC-CCAPL 2	60.05±0.12
CMC- CCAPL3	**35.08±0.16[*]**
CMC- CCAPR1[c]	125.02±0.2
CMC-CCAPR 2	80.09±0.15
CMC-CCAPR 3	**40.02±0.12[#]**

Values are represented as Mean ± SEM. n=3, *#Superscripts indicate significant differences in results ($p < 0.05$)

[a]CMC-GLY- Carboxymethyl cellulose + glycerol

[b] CMC-CCAPL- Carboxymethyl cellulose + cetyl caprylate ester

[c] CMC-CCAPR-Carboxymethyl cellulose + cetyl caprate ester

CONCLUSION

In summary-

- CCAPL and CCAPR both esters were successfully produced by the lipase catalyzed esterification reaction and highest yield of both the esters were obtained mainly by the NS435 lipase catalyzed reaction. Antimicrobial study was also confirmed by the well diffusion study.
- CMC-CCAPL or CMC-CCAPR composite films with improved water vapor permeability relative to the CMC films without CCAPL or CCAPR could be obtained by an emulsion technique.
- The main physicochemical properties of the CMC films were affected by glycerol, CCAPL and CCAPR. The increased concentration of glycerol decreased the mechanical strength and increased the water vapor permeability. CCAPL and CCAPR could reduce the water vapor permeability of the CMC films. Good thermal properties of films were also observed in all cases of CMC films containing glycerol, CCAPL and CCAPR.
- An appropriate proportion of glycerol and CCAPL or CCAPR is required to reduce water vapor transmission through CMC films, while having no unfavorable effect on the mechanical properties of the films.

CHAPTER 2

ENZYMATIC PREPARATION OF CASTOR OIL FATTY ACID BASED PLASTICIZER FOR INDUSTRIAL APPLICATION

INTRODUCTION

Esters of fatty acids and alcohols have multidimensional applications according to their molecular nature, which includes lubricants, greases, plasticizers, resins etc. The recent emerging interest of 'going green' even more increases the searching opportunities for use of fatty acid based derivatives because they possess low migration, low toxicity and better biodegradility. Castor oil and its derivatives have enormous opportunities to act as specialty chemicals in various purposes for the presence of ricinoleic acid, the 12- hydroxyl - 9-octadecenoic acid, the unique fatty acid with a pendant hydroxyl group and a double bond. Castor oil based plasticizers can be good substitute of phthalates or other plasticizers. It is odorless and completely biodegradable. It is reported that a plasticizer produced from castor oil makes plastic products soft and flexible and that is fully biodegradable. Furfuryl alcohol (FA) is a pentosan-rich biomass derived chemical obtained by the hydrogenation of furfural. FA is a biomolecule, which polymerizes readily under acidic conditions and forms a thermo-stable polymer that is widely applied in the foundry industry to produce biorenewable plasticizer. Most plasticizers are usually synthesized via esterification reactions between corresponding acids/anhydrides and alcohols in the presence of acidic catalysts. But implementation of lipase in the oleochemical industry has become lucrative because the conventional chemical process requires high energy cost and additional purification processes. Nowadays, introduction of lipase is a new edge in esterification reaction process which has a huge potential in industrial application. There are innumerable counts of studies regarding the biocatalytic esterification processes.

Poly (vinyl chloride) (PVC) is extensively used in daily life in its resilient form. It is used in construction materials, packaging materials, electronic materials and goods for everyday consumption -blood bags, urine bags, gloves, dialysis bags, blister packs for pharmaceutical usage and children toys. PVC contributes 70-90% of the total market in the majority of these application sectors. These PVC products need bulk quantity of plasticizers that could alter their thermal and mechanical properties. Presently, 500 kinds of plasticizers have been commercialized and most commonly used commercial plasticizers are petroleum based phthalate esters. According to the report of 2014, these esters of phthalate are accounted for 70% of the global plasticizer demand. There are numerous reports regarding the adverse effects of phthalates on the environment due to its migration from the polymer matrix during processing and this phenomenon causes the depletion of the service life of the polymer product leading to cause toxicity to the human health. In addition, the continuous diminution of fossil oils, striking fluctuations in the price of oil, there is a burning need to develop biodegradable and renewable resource based plasticizers to replace the phthalate based plasticizers. Vegetable oils are usually regarded as the most important class of renewable resources, because of their ready availability, innate biodegradability and low toxicity. This group includes soybean oil, sunflower oil, castor oil, linseed oil and their fatty acid esters. These kinds of vegetable oil based derivatives and their esters are also applied as plasticizer and additives to commercially available polymers for the improvement of their mechanical performances and biodegradability. But the preparation of vegetable oil based derivatives to substitute conventional petroleum based derivatives is a major modern-day challenge from both environmental and economical point of view.

The present chapter deals with the biocatalytic esterification of furfuryl alcohol (FA) and castor oil fatty acid (COFA) to produce a biodegradable, biorenewable and eco-friendly plasticizer. The chapter is again divided in two sections. Section A consists of the process parameter optimization of the esterification reaction along with characterization and performance evaluation of the ester as a biodegradable plasticizer and in Section B kinetic study of the enzymatic esterification reaction is presented.

Biocatalysis involves the implementation of natural enzymatic catalysts, such as lipases (e.g. RMIM, TLIM, NS435, *Candida rugosa*, pancreatic lipases, *Mucor japonicas*, *Rhizopus arrhizus*) instead of chemical catalysts. However, optimal immobilization of respective biocatalysts (here lipase) is fundamentally important for their enhanced usage, which ensure its smooth removal from reaction mixture for recycling purpose. Additionally, moderate reaction condition offered by lipase relative to chemical catalysis is a point of interest to avoid polymerization of FA and dehydration of COFA. Lipase B from *Candida antarctica* (NS 435), best known immobilized enzyme for esterification, was used for ester preparation.

Kinetic study of esterification reaction between FA and COFA using immobilized *Candida antartica* lipase B in a solvent free system was carried out using the well-known ping-pong bi bi mechanism. Subsequently, experimental conditions were optimized to maximize equilibrium yield of ester. Furthermore, different kinetic parameters, present in corresponding rate equation, such as the Michaelis-Menten constants, inhibition coefficients etc. were evaluated under varied conditions of temperature, molar ratio of reactants and enzyme loading in order to decipher the nature of their dependence on experimental parameters. Respective functional forms were quantitatively evaluated using standard Response Surface Methodology (RSM), which clearly marks the uniqueness of present work because respective outcome might be helpful in local optimization of process conditions in cases where the global optimum is beyond the scope of existing production facility.

OBJECTIVES

Objective of the work includes in Section A:

Biocatalytic preparation of ester of furfuryl alcohol (FA) and castor oil fatty acid (COFA) with immobilized lipase B from *Candida antarctica* (NS 435);

1. Characterization of ester by NMR, FTIR, fatty acid composition, viscosity and density etc.
2. Biodegradable property of the ester as natural plasticizer; and
3. Application of the prepared ester as plasticizer in poly vinyl chloride (PVC) and ethyl cellulose (EC) films and biodegradation study of the films.

Objective of the work includes in Section B:

1. Optimization of Parametric conditions to maximize the equilibrium yield of ester
2. Evaluation of Michaelis-Menten constant, inhibition coefficients and other important kinetic parameters.
3. Quantitative analysis of functional forms using Response Surface Methodology (RSM).

SECTION A

REVIEW OF LITERATURE

A major part of the polymer and packaging industry is dependent on plasticizer industry. For an example, a great part of the poly (vinyl chloride) is used as plasticized PVC. Hence, it is commonly used in cables, medical devices, and computers. Initially, most commonly used plasticizers were di-octyl phthalate (DOP), as di-n-butyl phthalate (DnBP), butyl benzyl phthalate (BBP), aliphatic di-carboxylic acids which are non-renewable resource based [90-91].Not only PVC material ,there are some selective biopolymeric materials such as ethyl cellulose (EC) , CMC, HPMC which are also compatible with these kinds of phthalate based plasticizers and have an enormous usage in pharmaceutical industries.EC is extensively used as coating agent and encapsulation in pharmaceuticals to protect core substances and attain sustainable drug release and also in the food, cosmetic, varnishes, lipsticks and insulation. Semi-synthetic cellulose derivative has applications as flavor fixative, food printing and cream thickening. Generally, commercially available phthalate based plasticizers are blended with these kinds of biopolymeric materials. Recent research in industrialized countries for toxins in human urine samples has uncovered the wide exposure of people to the group of phthalate plasticizers [91-95]. Phthalates are not bound chemically with the polymers and they can easily diffuse to the environment during its life time, particularly those with high fat content [96-98]. As a result, leaching of phthalates contaminate food products, drinks, and it belongs to the ubiquitous environmental contaminants and present difficulties in toys [93,94]. Several phthalates and especially di-ethyl hexyl phthalate (DEHP) known as di-octyl phthalate (DOP) are suspected of having carcinogenic and toxic effects [99-105]. In 2008, di butyl phthalate (DBP), DEHP, and benzyl butyl phthalate (BBP) were listed within the 14 SVHC (Substances of Very High Concern) by ECHA (European Chemicals Agency), as toxic for human infertility [106,150].

High toxicity of the commercially used petroleum based plasticizers, scarcity of petroleum resources and the rising cost of petroleum products has led to the development of alternate material sources for industrial uses. Nowadays, renewable resources have attracted increasing attention as non-petroleum raw materials to prepare chemical products [107]. The main chemical composition of the vegetable oil is triacylglycerol, and the main components of triglycerides are saturated and unsaturated fatty acids, which can be used as platform chemicals for many polymer and monomer synthesis. Vegetable oils are preferred by the chemical industry due to their world-wide availability and relatively low prices. The main chemical structure of vegetable oils is triacylglycerols [108,152,153], the main components of triglyceride are saturated and unsaturated fatty acids which could be used as platform chemicals for many polymers and monomers synthesis. Chemical products based on vegetable oil have been experienced fast development. Recently, the uses of vegetable oil as a substitute to petroleum-based raw materials for the designed and synthesized monomers have been receiving particular attention due to the environment and economic concerns. Reactive diluents were prepared from cardanol oil for epoxy resin [109-110]. Biodegradable multifunctional additive derived from soybean oil for ecofriendly lubricant was also prepared [112,113]. Biodegradable polyurethane stuffs have been synthesized from vegetable-oil-based polyol [114,115].

Basically triacylglycerols are major backbone of vegetable oils. The key components of triglyceride are saturated and unsaturated fatty acids, which have been used as base material for preparation of vegetable oil based plasticizer and experienced fast development [116,117,151,153].Current studies showed a trend of preparation of epoxidized vegetable oil as environmental plasticizer such as , epoxidized soybean oil [118-120], epoxidized sunflower oil [121], epoxidized safflower oil [122], epoxidized linseed oil [123], epoxidized cottonseed oil [124], and epoxidized cardanol diethyl phosphate [110]. According to a recent report, epoxidized soybean oil is used as commercial plasticizer for PVC materials because of its excellent water and oil resistance properties [118-120,154].Waste cooking oil was also used as bio based plasticizer for PVC films [125,134].Recent studies showed introduction of fatty acid derivatives as plasticizer into the polymeric materials. Oleic acid based bioplasticizer has been prepared for the PVC film formulation [131]. Another study showed the application of rice fatty acid based polyol esters as plasticizer into the polymeric materials[126]. Palm oil based plasticizer also showed good thermal stability over commercially available polymeric materials [127].These new plasticizers derived from renewable resources which are environmental friendly plasticizers.

In this respect, castor oil is a very potential candidate. Although castor oil is not edible, it is more versatile than other vegetable oils as it is widely used as a preliminary component for so many industrial products because of its unique chemical structure. Castor oil is one of those vegetable oils which have a potential application in chemical industries. It is a raw material for paints, coatings, inks, lubricants plasticizer and a wide variety of other products. It is a triacylglycerol containing one hydroxyl fatty acid, i.e. ricinoleic acid or 12- hydroxyl oleic acid that comprises over 89% of the fatty acid [128-130]. Castor oil is very much potent due to its unique molecular structure. There are three functional groups are present in this molecule, i.e. carboxylic groups, carbon double bonds (C=C) and pendant hydroxyl group [128], which make the oil capable for many chemical modifications. There are numerous reports on the preparation of various kinds of castor oil based plasticizers. PuyouJia [130] reported about the synthesis of castor oil based polyol ester for plasticization of PVC materials. Phosphorous containing castor oil derivative showed excellent flame retardant properties[129]. Epoxidized castor oil-based di-glycidyl-phthalate plasticizer also acted as a good thermal stabilizer[132].

Furfuryl alcohol (FA) is a pentosan-rich biomass derived chemical obtained by the hydrogenation of furfural. FA is a biomolecule, which polymerizes readily under acidic conditions and forms a thermo-stable polymer that is widely applied in the foundry industry to produce biorenewable plasticizer. According to a report, Oleic acid was successfully esterified with furfuryl alcohol [133].

The concept is to synthesize the ester of castor oil fatty acid and furfuryl alcohol biocatalytically. Most plasticizers are usually synthesized via esterification reactions between corresponding acids/anhydrides and alcohols in the presence of acidic catalysts[135].But implementation of lipase in the oleochemical industry has become lucrative because the conventional chemical process requires high energy cost and additional purification processes .Nowadays, introduction of lipase is a new frontier in esterification reaction process which have a huge potential in industrial application. There are innumerable counts of studies regarding the biocatalytic esterification process. Yadav and Devi [136] reported a kinetic study on the hydrolysis of tetrahydrofurfuryl butyrate by immobilized *Candida antarctica* lipase. There is an another study on the nylon immobilized lipase catalyzed esterification of fatty acids in n-hexane[137].Zhong et. al. reported a study on the lipase catalyzed synthesis of alkyl oleates in solvent free systems[138].Shintre, Ghadge and Sawant esterified lauric acid with cetyl alcohol by immobilized lipase[139].D-isoascorbyl palmitate also synthesized with lipase catalyzed reaction esterification reaction[140]. Sengupta et al. [133] synthesized ester of furfuryl alcohol and oleic acid by immobilized *Candida antarctica* lipase B.

MATERIALS AND METHODS

LIPASE CATALYZED ESTERIFICATION REACTION OF FURFURYL ALCOHOL AND CASTOR OIL FATTY ACID

(i)Materials

Pure castor oil fatty acids (COFA)was obtained from Aloka Oil Industry, Kolkata and analyzed at laboratory for fatty acid composition by gas chromatography (GC). Furfuryl alcohol (FA) was purchased from Merck India Ltd., Mumbai, India. NS 435, *Candida antarctica* lipase B, immobilized using macroporous acrylic resin with a water content of 10 % (w/w), a product of Novozyme India Pvt. Ltd., was obtained from Brenntag Ingredients (India) Pvt.Ltd., Mumbai, India.

(ii) Methods

For the esterification reactions conducted with stoichiometric molar ratio of FA (MW 98.10 g/mol) and COFA (MW 298.4 g/mol), were taken in different molar ratios (32.87g of FA and 100g of COFA for 1:1 molar ratio,98.61g of FA and 100g of COFA for 3:1molar ratio,164.35g of FA and 100g of COFA for 5:1 molar ratio) in a flat bottom flask and stirred at 200 rpm at different temperatures(45^{0}C to 65^{0}C) for different course of time with changing amounts (4% to 12% w/w) of immobilized lipase. The reaction was performed in a complete solvent free environment and under vacuum (Vacuum Pressure 28mmHg). The substrate molar ratio, temperature and catalyst loading of the immobilized lipase were variable experimental parameters and optimized with a single reaction mixture that were studied one at a time and eventually followed in the subsequent reactions.

To study the effect of the substrate molar ratio on the ester yield, the molar ratio of substrates was altered as 1:1, 3:1 and 5:1 of FA and COFA. The temperature and immobilized lipase concentration were kept constant at 60^0C and 10% lipase [w/w, of the previously mentioned total substrate weight] respectively at different time intervals (0 h to 6h). The reverse molar ratios i.e. excess amount of COFA were not tried as the removal of excess fatty acid will be difficult from the ester of FA and COFA.

The influence of temperature on the rate of reaction of FA and COFA was studied by conducting the reactions at different temperatures ranging from 40^0C to 65^0C, maintaining other reaction parameters such as the substrate molar ratio (FA:COFA 3:1) ,concentration of immobilized lipase (10% w/w, of the total substrate weight) and reaction time (5h) constant.

The difference in immobilized lipase concentration ranging from 4%to 12% (w/w, of the total substrate weight) was also studied by maintaining the other reaction parameters such as temperature (60^0C) ,substrate molar ratio (FA: COFA 3:1) and the reaction time (5h) constant. The yield of the product was expressed as g of ester present in 100g of the product mixture (%w/w)[**Scheme 3**].

The reaction period was monitored by determining the intermittent acid value of the product during the course of time. The ester formation was confirmed by thin layer chromatography (TLC), and the percent yield of ester and fatty acid composition of the ester was determined simultaneously.

The final product was distilled by vacuum distillation under low pressure (Vacuum Pressure 28mmHg) at 90^0C to remove excess FA and finally washed with dilute NaOH solution followed by water to remove free fatty acids. The product was dried under vacuum (Vacuum Pressure 28mmHg) and stored for further analysis.

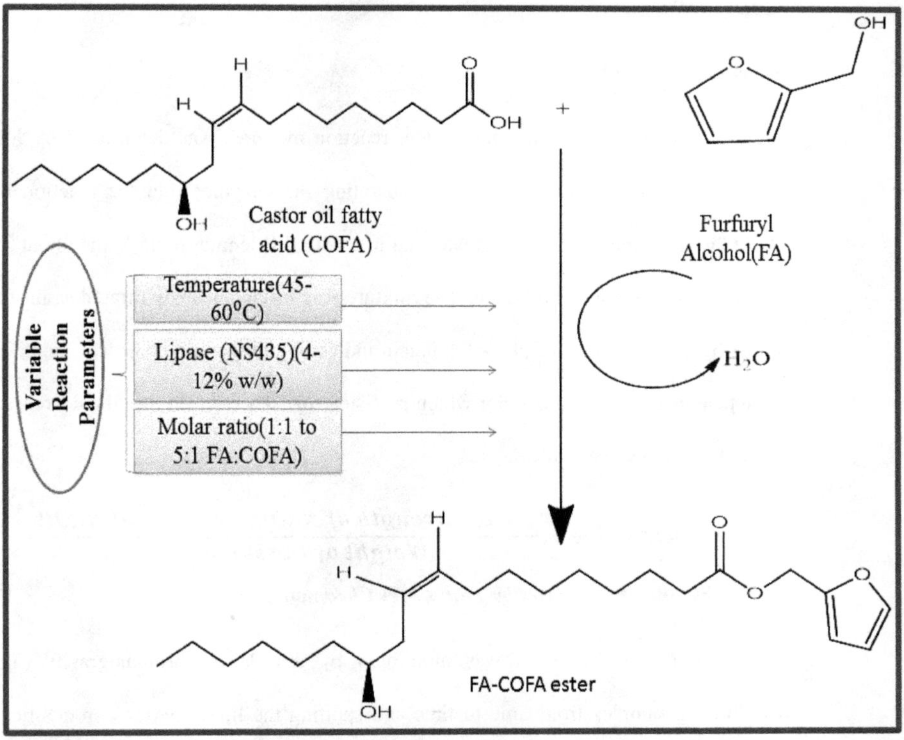

Scheme 3. Schematic diagram of enzymatic esterification reaction of castor oil fatty acid (COFA) and furfuryl alcohol (FA)

PHYSICOCHEMICAL CHARACTERIZATION OF THE ESTER PRODUCED BY THE ESTERIFICATION OF FURFURYL ALCOHOL (FA) AND CASTOR OIL FATTY ACID (COFA)

Determination of acid value:

Acid value of the initial and intermittent reaction mixtures was determined by following standard method of AOCS [84] after separating the enzyme from the reaction mixture. About 2g of reaction mixture was weighed in a 250ml Erlenmeyer flask and about 25 ml of hot neutralized alcohol was added. The mixture was instantaneously titrated against 0.1 (N) NaOH by adding 0.5 ml of phenolphthalein indicator and the end point was identified with the appearance of light pink color which persisted for 30 seconds. From the decrease of acid value yield of ester was calculated.

$$\boldsymbol{Acid\ value} = \frac{\boldsymbol{56.1 \times Strength\ of\ NaOH \times Volume\ of\ NaOH}}{\boldsymbol{Weight\ of\ the\ sample\ oil}}$$

Qualitative analysis of esters by Thin Layer Chromatography (TLC):

Production of COFA esters was monitored by thin layer chromatography (TLC) by withdrawing samples from time to time and spotting the lipid mixtures on a Silica Gel G plate (0.2 mm thick). Here hexane-diethyl ether (70:30) was used as the developing solvent and different bands were visualized on iodine absorption and identified by measuring R_f value [85].Pure castor oil and COFA were used as the standards.

Analysis of fatty acid composition of COFA and its esters by GC:

The ester produced was first hydrolyzed with alkali and then the liberated fatty acids were methylated with BF_3 in methanol and subjected to gas chromatographic (GC) analysis [86]. The GC (Agilent 6890 N) was fitted with a DB-Wax capillary column (30 m x 0.32 mm x 0.25 μm) and a flame ionization detector. The rate of N_2, H_2 and airflow was kept at 1, 30 and 300 ml/min, respectively. The inlet and detector temperatures were kept at 250°C and the oven temperature was programmed to increase from 150°C to 190°C at a rate of 15°C/min, then to hold for 5 min, and then to increase to 230°C at a rate of 4°C/min, and then to hold for 20 min.

Fourier Transform Infrared Spectroscopy (FTIR) of esters:

FTIR analysis has been carried out to confirm the ester linkage and presence of free hydroxyl group. A PerkinElmer make Spectrum GX spectrophotometer with 400–4000 cm^{-1} range was used for the analysis. A very thin film of products covered the NaCl cells (25mmi.d × 4mmthickness) was prepared and was used for analysis [141].

^{1}H NMR and ^{13}C NMR:

^{1}H and ^{13}C nuclear magnetic resonance spectroscopy (^{1}H NMR and ^{13}C NMR) were performed under ambient temperature on a 300MHz Brucker spectrometer using $CdCl_3$ as the corresponding solvent [142].

Physical parameters:

Physical parameters such as flash point, pour point, viscosity and density were determined by using ASTM methods [ASTM D56-05, ASTM D 97-96a, ASTM D 445-04C2, ASTM D891-2].

BIODEGRADATION STUDY OF THE ESTER

(i) Materials

Microorganism:

Ochrobactrum pseudintermedium, used in this study was previously isolated from soil of iron ore handling mine area of Durgapur, West Bengal, which was contaminated with machine lubricating oil and was identified by 16s rRNA (gene accession number KJ094035). This strain was reported to degrade lubricating oils.

Media:

The medium used for the growth of the bacteria was Bushnell Haas mineral salt medium (BH medium) (Hi media, India) which was recommended for studying hydrocarbon deterioration. This medium contained gram per litre (g/l) of distilled water $MgSO_4$ - 0.2, $CaCl_2$ - 0.02, KH_2PO_4 - 1, K_2HPO_4 - 1, NH_4NO_3 - 1 and $FeCl_3$ - 0.05.

(ii) Method

Biodegradation of the FA-COFA ester (ester of FA and COFA) was a five day study. *Ochrobactrum pseudintermedium* were grown in culture tubes, in sterile BH medium [143]. Ester of FA and COFA was the sole carbon source of the BH medium and the ester was added to the medium at different weight percentages i.e. 1%, 2%, 3%, 4% and 5%(w/v) basis (e.g.0.1g of FA-COFA ester dissolved in 10 ml of BH media). The media containing different concentrations of ester were incubated for five days at 37^{0}C for the growth of the microorganism. Absorbance of each media, containing different concentrations of ester was measured at 600nm using Shimadzu (1601A) UV-VIS Spectrophotometer to observe the growth of the microorganism. For each set of FA-COFA ester containing medium (expressed in percentages) a control set (e.g. media containing 1 %(w/v), have a separate control set against an experimental set) , was prepared without addition of bacterial inocula and the methods were repeated thrice for statistical analysis.

A gravimetric study of the reaction product was done for each sample to observe the biodegradability of the product. It means that, the FA-COFA ester was added to the medium at different weight percentages i.e. 1%, 2%, 3%, 4% and 5% basis (e.g.0.1g of FA-COFA ester dissolved in 10 ml of media) and the media was incubated for five days. After the completion of the five days study, the FA-COFA ester was extracted from the media with chloroform, dried and weighed to observe degradation of the ester by comparing the data with the initial weight of the ester added to the medium. For each set of FA-COFA ester containing medium (expressed in percentages) a control set (e.g. media containing 1 %(w/v), have a separate control set against an experimental set) , was prepared and the methods were repeated thrice for statistical analysis.

PREPARATION AND PERFORMANCE EVALUATION OF THE FA-COFA ESTER PLASTICIZED PVC FILMS

Preparation of the PVC film:

(i) Materials

Polyvinyl Chloride (PVC) resin was a gift from Auropol India Pvt.Ltd., Stearic acid, Zinc Stearate and Tetrahydrofuran and all other chemicals were analytical grade reagent from Merck. Di-butyl phthalate (DBP) was procured from Loba Chemie Pvt.Ltd., Mumbai, India.

(ii) Method

Plasticization of PVC by FA-COFA ester as plasticizers was conducted using solvent–cast film method. The FA-COFA ester based plasticizers were added as substitute of DBP at the level of 20, 40, 60, and 80% of total plasticizers used [127]. The chemical compositions in formulation of PVC resin are shown in the following table[Table 22]. The plasticizers (mixture of DBP and FA-COFA ester based plasticizer) and PVC resin (added with zinc stearate and stearic acid) were dissolved in 30 ml of tetrahydrofuran (THF) at 40^0C. Further, the solution was casted into a film, and then it was kept for 12 h and stored in vacuum oven for 24 h at 40^0C before it was measured for further performance evaluation**[Scheme 4].**

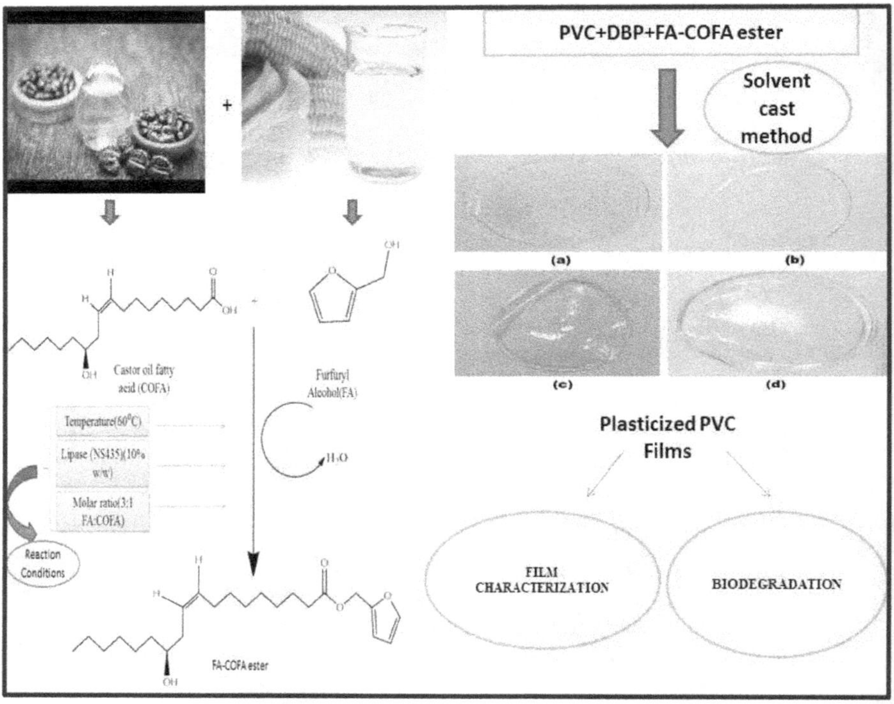

Scheme 4. Application of FA-COFA ester in PVC Films

Table 22. Chemical composition in the formulation of PVC resins

Components	Amount(g)
Plasticizer	1.35(60 Phr)
PVC	2.25
Zn-stearate	0.0450(2 phr)
Stearic acid	0.0068(0.3 phr)
TOTAL	3.65

*phr =per hundred resin.

Film Characterization:

X-Ray diffraction:

In order to observe alterations of crystallinity of the PVC films induced by FA-COFA ester and DBP as plasticizer addition, X-raydiffraction was performed on film samples with a P.W 3040/60model X-ray diffractometer PAN analylitical (Netherland) with. The X-ray source was operated at 40 kV and 30 mA. Angular scans from 5.02 to $99.97^0(2\Theta)$ were performed at a $2^0C/min$ scan rate with a Cu source X-ray tube.

Scanning electron microscopy (SEM):

A thin layer of Pt was sputter-coated with PVC film samples to study the SEM images. Afterwards, the coated samples were studied by Zeiss EVO 18 Special Edition (Germany) with 15 kV accelerating voltage in secondary electron mode. The magnification of SEM images was 10 kx.

Mechanical properties:

Tensile strength, elongation at break and elastic modulus were performed by using a universal instrument of mechanical properties (Instron, 5565, USA) with a 50 N loaded cell with a constant extension speed of 0.1 cm/s. Rectangular shaped films with a length of 50 mm and a width of 15 mm, were clipped by two clamps on instrument. Elastic modulus was calculated from the slope of the stress and strain curves within the range of elastic limit of stretching. The tensile strength was expressed as the maximum stress at break divided by the initial cross-sectional area of the films and the elongation at break as a percentage of the original length.

Thermogravimetric analysis (TG) and Differential thermal analysis (DTA):

TG was performed in an aluminum pan by Diamond TG/DTA thermal analyzer from Perkin Elmer, using Pyris software, Flow rate of Nitrogen of 25 ml/min was used as the purging gas. The heating range of samples was from 30°C to 650°C with a rate of 10°C/min. DTA was determined by using a Diamond TG/DTA thermal analyzer from Perkin Elmer, using Pyris software. About 5 mg of the film sample was weighed and scanned under nitrogen flow at a rate of 25 ml/min. Film sample was preheated to 150°C at the initial phase at a rate of 20°C/min for the removal of water and elimination of thermal history. After cooling to 20°C, the sample was then heated to 700°C in a second scan at a rate of 10°C/min.

Migration stability tests:

Test for migration stability of FA-COFA ester and DBP combined PVC film samples were performed using three different solvents such as petroleum ether [126], gasoline and water at 25°C. This test was performed amongst pure PVC film, PVC+(DBP: FA-COFA :: 8:2), PVC+(DBP: FA-COFA :: 6:4), PVC+(DBP: FA-COFA:: 4:6) and PVC+(DBP: FA-COFA :: 2:8). PVC film samples containing DBP and FA-COFA ester were weighed thrice and kept in 100 ml of each solvent. Those film samples were taken out after 72 h and dried in a vacuum oven at 30°C for 24 h. The weights of the dried samples were determined.

Film degradation:

Soil Biodegradation studies:

The soil used was obtained from the Ganges river bank, Kolkata. The soil was slightly acidic with a pH value of 6.5. The blend films were embedded in soil which was taken in a beaker. The soil moisture content was about 30% and it was maintained throughout the course of study. For every individual type of the PVC blend films, three test sets prepared separately for soil degradation test and loss of weight was carefully determined and averaged. The beakers were set aside at a well-ventilated place by maintaining an average temperature of 30^0C.

Enzymatic degradation:

In the present study, PVC blend films were kept for degradation process in a buffer solution containing lipase enzyme from porcine pancreas [123]. Known amounts of Na_2HPO_4 and KH_2PO_4 solution were added for the preparation of phosphate buffer solutions of pH 7.5. Enzyme containing buffer solution was prepared by adding a certain amount of lipase and the concentration used was 250 units per 10 ml. Those PVC blend films were weighed and kept immersed in enzyme solution at room temperature (30^0C).During the study period, the films were removed after a certain time intervals (10, 20 and 30 days).Films were washed thoroughly with water and dried in hot air oven and weighed.

Hydrolytic degradation in buffer solutions:

The PVC blend films were weighed and held in phosphate buffer solution of pH 7.5 in a glass container at room temperature (30^0C) [123]. Phosphate buffer solution was prepared by the combination of KH_2PO_4 and Na_2HPO_4 solutions. After the time intervals of 10, 20 and 30 days, the films were removed from the test solution and washed thoroughly with water. Then the films were dried and weighed. The test was repeated thrice for statistical significance.

PREPARATION AND PERFORMANCE EVALUATION OF THE FA-COFA ESTER PLASTICIZED ETHYL CELLULOSE (EC) FILMS

(a)Preparation of the EC film:

(i)Materials

Ethyl cellulose was procured from Hi-media Laboratories Pvt. Ltd. of CP grade with 48–49.5% ethoxy content. All analytical grade chemicals were used and obtained from Merck, India Ltd, Mumbai, India.

(ii) Method

Each film sample was prepared by following the method of Yang et al[112] and the method was slightly modified. In each film sample, 20 ml of acetone was blended with a certain amount of EC at 30^0C and the mixture was stirred rigorously by magnetic stirrer. FA-COFA ester and DBP were added as plasticizer into the EC solution separately and the solution was stirred at 30^0C for1 h. The plasticizer concentration was varied in a range of 5-25%.Such as; EC + FA-COFA 15 represent 85% EC (1.7 g) and 15% FA-COFA ester (0.3 g) (Table 44). Subsequently, the solution was cast into a clean glass petridish (diameter 14 cm) and immediately it was covered with a clean glass cover to control the rate of evaporation of acetone. Films were dried in an oven at 40°C for 6 h. Thereafter, the film was peeled off and stored at a temperature of 23 ± 0.5^0C. Films made by casting were reproducible and compared with the controls **[Scheme 5].**

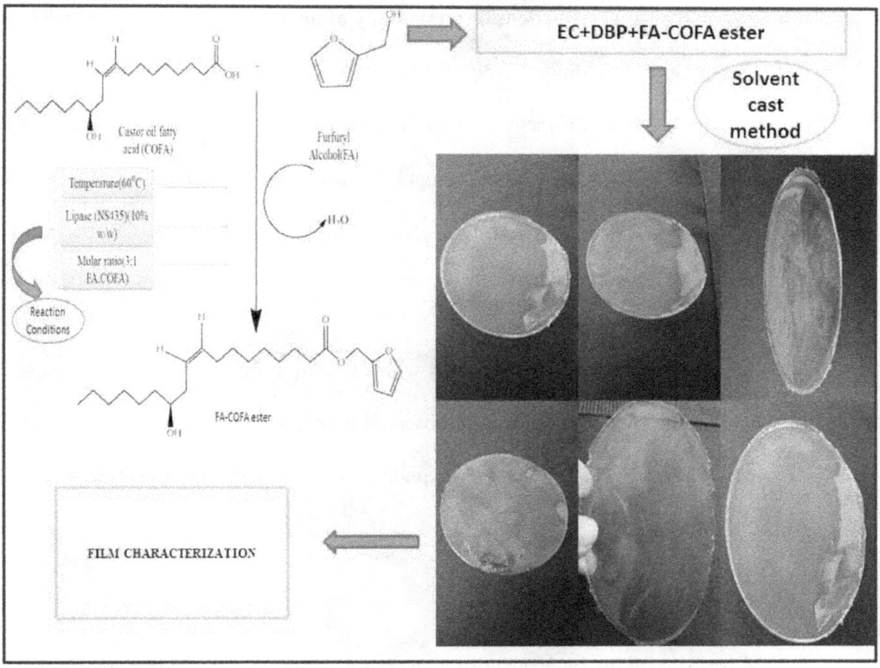

Scheme 5.Application of FA-COFA ester in EC Films

(b) Film Characterization:

X-Ray diffraction:

In respect to observe alterations of crystallinity of the films induced by FA-COFA ester and DBP as plasticizer addition, X-ray diffraction was performed on film samples with a P.W 3040/60model X-ray diffractometer PAN anaylitical (Netherland) with. The X-ray source was operated at 40 kV and 30 mA. Angular scans from5.02 to $99.97^0(2\Theta)$ were performed at a 2^0C/min scan rate with a Cu source X-ray tube.

Scanning electron microscopy (SEM):

A thin layer of Au was sputter-coated with EC film samples to observe the SEM images. Thereafter, the coated samples were observed by Zeiss EVO 18 Special Edition (Germany) with 15 kV accelerating voltage in secondary electron mode. The magnification was 10 kx in SEM images.

Atomic force microscope (AFM):

AFM was carried out on Veeco di Innova (Bruker AXS Pvt. Ltd.,USA) by placing the film samples onto the mica surface. All of the images were recorded in tapping mode by silicon cantilevers. Commercial probes were used with a spring constant of 20–80 N/m and a resonance frequency of 240–308 kHz. The scale in testing was 10.0 m × 10.0 m. The software Nanoscope V1.40 was used to process the AFM images.

Mechanical properties:

A universal instrument of mechanical properties (Instron,5565,USA) with a 50 N loaded cell was used for tensile strength, elongation at break and elastic modulus testing in a constant extension speed of 4 mm/min. Films were cut into rectangular shape with a length of 50 mm and a width of 15 mm, and were then clipped by two clamps on instrument. The elastic modulus was calculated from the slope of the stress–strain curves within the range of elastic limit of stretching.

Thermogravimetric analysis (TG) and differential scanning calorimetry (DSC):

TG was performed in an aluminum pan by TGA/DSC 1 thermal analyzer from Mettler, Toledo, Switzerland, using software StareSW9.10. Flow rate of Nitrogen of 25 ml/min was used as the purging gas. The heating range of samples was from 30^0C to 700^0C with a rate of $10^0C/min$.

DSC was determined by using a TGA/DSC 1 thermal analyzer from Mettler, Toledo, Switzerland, using the software Star SW 9.10.About 5 mg of the film sample was weighed and scanned under nitrogen flow at a rate of 25 ml/min. Film sample was preheated to 150^0C at the initial phase at a rate of $20^0C/min$ for the removal of water and elimination of thermal history. After cooling to 20^0C, the sample was then heated to 220^0Cin a second scan at a rate of $10^0C/min$.

Water vapor permeability:

Water vapor permeability of the EC films with or without plasticizer (FA-COFA ester and DBP) was calculated in accordance with the modified ASTM E96-00 method. Film samples were sealed over a 60 mm circular opening of a permeable container with calcium chloride (0% RH inside the cell). Then the permeable containers were placed inside a desiccator containing the saturated sodium chloride solution (75% RH outside the cell), to create a 75% RH gradient across the film [87]. The water vapor transport was determined from the weight gain of the permeable container at an interval of every 24 h until the stable weight was attained. Water vapor transmission rate was calculated by using the following equation [88]:

$$Q = WL/S$$

where W is increase in the desiccant weight per 24 h, L is the film thickness (cm), S is the exposed surface area (cm^2) and Q is the water vapor transmission rate ($g/cm^2/24$ h).

Migration stability tests:

Migration stability tests of the FA-COFA ester and DBP from film samples was performed using three different solvents such as petroleum ether [126], gasoline and water at 25°C. Migration stability test was performed between EC film containing 15% of FA-COFA ester and the EC film with 15% of DBP. EC film samples containing FA-COFA ester and DBP were weighed thrice and kept in 100 ml of each solvent. Those film samples were taken out after 72 h and dried in a vacuum oven at 30°C for 24 h. The weights of the dried samples were determined.

The percentage weight loss was calculated in relation to the initial weight of FA-COFA ester and DBP added to the EC film.

STATISTICAL ANALYSIS

All the analysis was done in triplicates (n = 3). All the data are represented as means with their standard errors (Mean ± SEM). Statistical comparisons among groups were performed using one way ANOVA. Level of significance $p < 0.05$ was considered by Tukey test (using Origin Pro 8 software).

RESULT AND DISCUSSION

OPTIMIZATION OF REACTION PARAMETERS

Change in molar ratio and temperature:

The progress of esterification reaction could easily be monitored by change in acid value of the reaction mixture. With the formation of ester, the acid value should decrease due to the loss of free fatty acid from the reaction mixture. **Table 23 and Table 24 (vide Figure 15a and Figure 15 b)** showed the change in acid value of the reaction mixture with time in different molar ratios and temperature respectively, keeping the other parameters constant. From Figure15 (a), the acid values at 0 h to 6 h for 1:1, 3:1 and 5:1molar ratios of FA and COFA were determined using NS 435 as biocatalyst (10% w/w) at 60^0C.In case of 3:1 molar ratio of FA and COFA, Figure15 (a) showed that there was a decrease in acid value with time and the highest decrease in acid value was obtained at the 5 h of the reaction. But an increase in the acid value was observed in the 6 h of the reaction time. This pattern of the reaction was also observed in other molar ratios (e.g.1:1 and 5:1) of FA and COFA at 0 to 6 h reaction time. Hence, from the Figure15(a) it was shown that a significant ($p<0.05$) decrease in acid value was observed at the 3:1 molar ratio of FA and COFA at 5h of the reaction time. From the Figure15(b) acid values of 3:1 molar ratio of FA and COFA at different temperatures ranging from 45^0C to 65^0C were determined at a constant time (5h) using immobilized *Candida antarctica* Lipase B as biocatalyst (10% w/w).In case of 60^0C temperature, Figure 15 (b) showed the highest decrease of acid value in comparison to other reaction temperatures (e.g. 45^0C, 50^0C, 55^0C and 65^0C).When the acid values at different temperatures were compared among each other, Figure.15(b) showed that the acid value at 60^0C was significantly ($p<0.05$) decreased in comparison to 45^0C, 50^0C and 65^0C. But acid value was not significantly ($p<0.05$) decreased in case of 55^0C when it was compared with 60^0C.Though the highest decrease in acid value was observed, therefore, from the Figure

15(b) a significant ($p<0.05$) decrease of acid value was obtained at 60^{0}C reaction temperature of FA and COFA at 3:1 molar ratio using 10% (w/w) immobilized *Candida antarctica* Lipase B as biocatalyst. Therefore this part of the study was concerned with different molar ratio of FA and COFA (1:1,3:1 and 5:1) and different reaction temperature. From Figures 15 (a) and 15(b), it could be said that 3:1 molar ratio of FA and COFA was the optimal molar ratio and 60^{0}C was the optimal temperature especially for the lipase catalysis.

Table 23.Changes in acid value (a) at different molar ratios (1:1,3:1 and 5:1) of FA and COFA using NS 435 as biocatalyst (10%w/w) at different time intervals (0 h to 6 h) at 60°C

Time(h) ↓	**Acid Value**		
	Molar ratio(FA:COFA)		
	1:1	**3:1**	**5:1**
0	87.20±2.85	87.96±2.40	86.89±1.07
1	31.31±0.88	16.32±1.43	20.34±0.5
2	17.13±0.32	11.46±1.14	14.2±0.47
3	14.57±0.56	9.26±0.67	12.23±0.28
4	9.86±0.32	5.09±0.70	7.78±0.07
5	8.91±0.05	**1.02±0.36***	4.52±0.5
6	10.60±0.30	5.65±0.56	8.06±0.20

Values are mean ± SEM. $p<0.05$, n=3

Table 24.Changes in acid value at different reaction temperature (45^{0}C to 65^{0}C) using NS 435 as biocatalyst (10% w/w) at a constant molar ratio (3:1) of FA and COFA and at a constant reaction time (5h)

Temperature (^{0}C)	**Acid Value**
45	3.6±0.08***
50	2.98±0.01**
55	1.13±0.03*
60	**1.02±0.01**
65	2.61±0.06$^\#$

Values are mean ± SEM. $p<0.05$, n=3

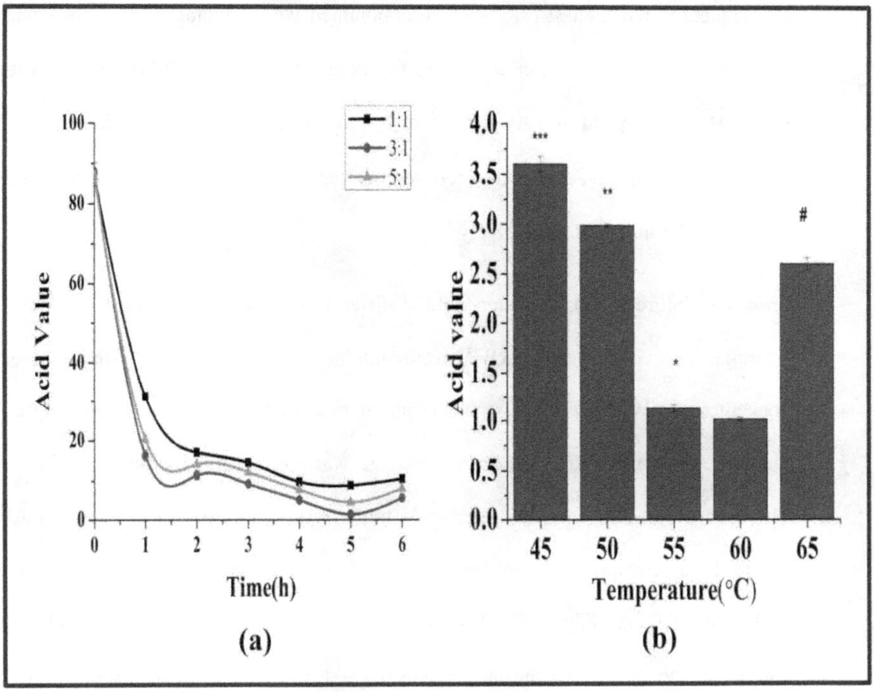

Figure 15.Changes in acid value (a) at different molar ratios (1:1,3:1 and 5:1) of FA and COFA using NS 435 as biocatalyst (10%w/w) at different time intervals (0 h to 6 h) at 60°C (b) at different reaction temperature (45°C to 65°C) using NS 435 as biocatalyst (10% w/w) at a constant molar ratio (3:1) of FA and COFA and at a constant reaction time (5h)

(a)Values are mean ± SEM, $p<0.05$, n=3.

(b)Values are mean ± SEM, $p<0.05$, n=3.

Statistical analysis:

*Significant at < 0.05 in temperature 60°C vs 55°C

**Significant at <0.05 in temperature 60°Cvs 50°C

***Significant at <0.05 in temperature 60°vs 45°C

#Significant at <0.05 in temperature 60°C vs 65°C

Consequently, the highest ester yield was obtained with 3:1 molar ratio of FA and COFA and at 60°C. This was further confirmed from the **Table 25 and Table 26(vide Figures 16 a and 16 b)**.The yield of ester with time (0 h to 6 h) at different molar ratios (1:1, 3:1 and 5:1) of FA and COFA and at different temperatures (45^{0}C to 65^{0}C) was plotted in Figures 16 (a) and Figure 16(b) respectively.

From the Figure.16 (a), the ester yield at different time scale (0 h to 6 h) for 1:1, 3:1 and 5:1molar ratio of FA and COFA were obtained using 10% (w/w) immobilized lipase concentration. With the increase in the molar ratio of FA and COFA from 1:1 to 3:1, the yield of ester was found to increase; a decrease was noticed in 5:1 molar ratio (Figure. 16 a). The decrease in the ester yield for the 5:1molar ratio of FA and COFA was quite obvious due to the inactivation of lipase because of higher amount of FA. In case of 3:1 molar ratio of FA and COFA, Figure16 (a) showed significantly ($p<0.05$) highest yield of ester in comparison to the rest of the molar ratio (e.g.1:1 and 5:1) at 5h of reaction time at 60^{0}C temperature and at 10% (w/w) immobilized lipase concentration. But at 6h of the reaction time, there was decrease in ester yield at all the molar ratio (1:1, 3:1 and 5:1) of FA and COFA .The decrease of the yield of ester at 6 h might be due to the polymerization of FA. From Figure. 16 (b), the ester yields of 3:1 molar ratio of FA and COFA at different temperatures ranging from 45^{0}C to 65^{0}C were determined at a constant time (5h) using immobilized *Candida antarctica* lipase B as biocatalyst 10% (w/w). At the 60^{0}C temperature, Figure.16 (b)showed the highest increase of ester yield in comparison to other reaction temperature (e.g. 45^{0}C,50^{0}C,55^{0}C and 65^{0}C).When the ester yield of different temperatures were compared among each other, Figure.16 (b) showed that the ester yield at 60^{0}C was significantly($p<0.05$) increased in comparison to 45^{0}C,50^{0}C and 65^{0}C. But the ester yield was not significantly ($p<0.05$) increased in case of 55^{0}C when it was compared with 60°C. With the increase in temperature from 45°C up to 60°C, there was an increase in the ester yield. A significant ($p<0.05$) decrease in the yield of ester was observed at 65°C (Figure.16 b).This decrease might be due to the inactivation of lipase at higher temperature as they are efficiently active at 60°C temperature. Therefore, the molar ratio of 3:1 of FA and COFA could be regarded as optimum substrate ratio due to its maximum yield on the basis of COFA at the end of the reaction and the apparent optimum reaction temperature could be 60°C.

Table 25.Changes in ester yield (% w/w) (a) with time at different molar ratios (1:1,3:1 and 5:1)of FA and COFA using NS 435 as biocatalyst (10%w/w) at different time intervals (0h to 6 h) 60°C

Time(h) ↓	Yield of Ester (%w/w) Molar ratio (FA:COFA)		
	1:1	3:1	5:1
0	0	0	0
1	45.65±0.58	56.02±2.65	30.13±0.07
2	53.44±0.57	64.72±2.53	59.24±0.16
3	62.27±0.75	68.83±0.80	65.79±0.11
4	69.50±0.53	78.04±1.23	72.12±0.04
5	79.11±0.65	**88.94±1.87***	82.45±0.05
6	68.38±0.71	77.42±1.74	72.23±0.74

Values are mean ± SEM. $p<0.05$, $n=3$

Table 26.Changes in ester yield at different temperature (45°C to 65°C) using NS 435 as biocatalyst (10%w/w) at a constant molar ratio (3:1) of FA and COFA at a constant reaction time (5h)

Temperature(°C)	Yield of Ester (%w/w)
45	51.06±0.80***
50	62.11±0.81**
55	76.05±0.10*
60	**88.94±0.82**
65	65.55±0.66#

Values are mean ± SEM. $p<0.05$, $n=3$

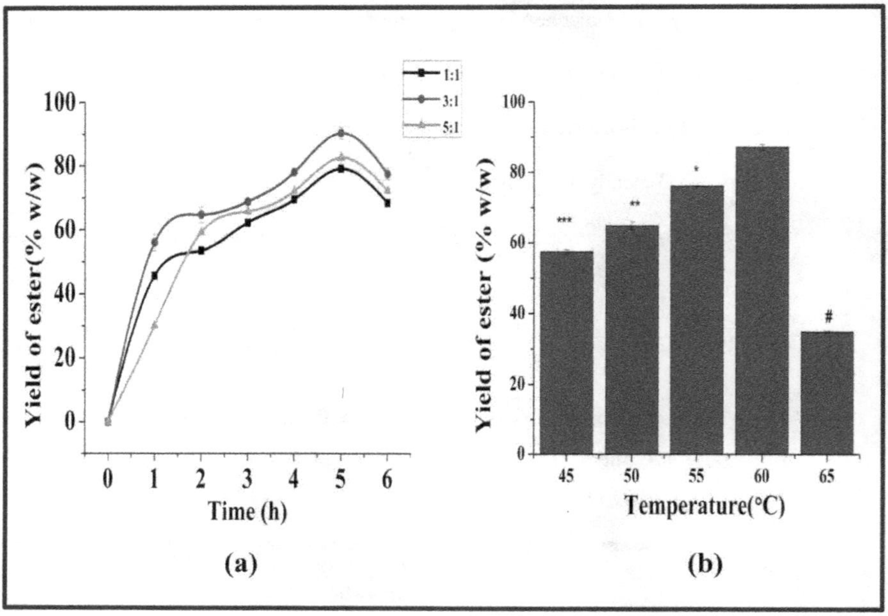

Figure 16.Changes in ester yield (% w/w) (a) with time at different molar ratios (1:1,3:1 and 5:1)of FA and COFA using NS 435 as biocatalyst (10%w/w) at different time intervals (0h to 6 h) 60^{0}C (b) at different temperature (45^{0}C to 65^{0}C) using NS 435 as biocatalyst (10%w/w) at a constant molar ratio (3:1) of FA and COFA at a constant reaction time (5h)

(a) Values are mean ± SEM, $P<0.05$, n=3.

(b) Values are mean ± SEM, $P<0.05$, n=3.

Statistical analysis:

*Significant at < 0.05 in temperature 60^{0}C vs 55^{0}C

**Significant at <0.05 in temperature 60^{0} vs 50^{0}C

***Significant at <0.05 in temperature 60^{0} vs 45^{0}C

#Significant at <0.05 in temperature 60^{0}C vs 65^{0}C

Change in immobilized lipase concentration:

The impact of immobilized lipase concentration was assessed with a constant substrate molar ratio of FA and COFA (3:1), time (5h) and temperature of reaction (60^{0}C). On the basis of total substrate weight 4%, 6%, 8%, 10% and 12% (w/w) concentrations of immobilized lipase were considered. From the **Table 27 and Table 28 (vide Figures 17a and 17b)**, it was observed that, with the increase in immobilized lipase concentration there was decreasing acid value and increased ester yield. From Figure 17 (a), the acid values of 3:1 molar ratio of FA and COFA at different immobilized lipase concentration ranging from 4% to 12% (w/w) on the basis of total substrate weight were determined at a constant time (5h) at 60^{0}C.In case of 10% (w/w) immobilized lipase concentration, Figure 17 (a) exhibited the highest decrease in acid value in comparison to other concentrations of lipases (e.g. 4%,6%,8% and 12% w/w).When the acid values of different lipase concentration were compared among each other, Figure 17(a) revealed that the acid value at 10%(w/w) immobilized lipase concentration was significantly($p<0.05$) decreased in comparison to 4%, 6% and 8% (w/w) immobilized lipase concentrations. But in case of 12% (w/w) immobilized lipase concentration, the acid value was slightly increased when it was compared to 10% (w/w) immobilized lipase concentration. A slight increase in the acid value was observed at 12% (w/w) immobilized lipase concentration which might be due to the enzyme substrate inhibition effect. Figure 17 (b) depicted the ester yield of 3:1 molar ratio of FA and COFA at different immobilized lipase concentrations ranging from 4% to12% (w/w) at a constant temperature (60^{0}C).At 10% (w/w) immobilized lipase concentration, Figure 17 (b)exhibited the highest ester yield in comparison to other immobilized lipase concentrations (e.g. 4%, 6%, 8% and 12% w/w).When the ester yields of different immobilized lipase concentration were compared among each other, Figure 17(b) displayed that the ester yield at 10% (w/w) immobilized lipase concentration was significantly($p<0.05$) increased in comparison to 4%, 6% and 8% (w/w) immobilized lipase concentration. But in case of 12% (w/w) immobilized lipase concentration, the yield of ester was slightly decreased when compared with 10% (w/w) immobilized lipase concentration. Hence, the optimum immobilized lipase concentration was noticed at 10% (w/w), because higher concentration led to a decrease in the ester yield.

Table 27.Effect of lipase (NS 435) concentration (4% to 12% w/w) (a) on acid value of the reaction of FA and COFA at 3:1 molar ratio and at 60^0 C temperature at a constant time (5h)

Lipase Concentration(%w/w)	Acid Value
4	4.67±0.31***
6	3.11±0.01**
8	2.8±0.03*
10	**1.02±0.03**
12	1.35±0.02$^{\#}$

Values are mean ± SEM. $p<0.05$, n=3

Table 28.Effect of lipase concentration on the yield of ester of the reaction of FA and COFA at 3:1 molar ratio and at 60^{0}C temperature at a constant time (5h)

Lipase Concentration(%w/w)	Yield of Ester(%w/w)
4	58.55±0.6***
6	71.43±0.36**
8	76.53±0.24*
10	**88.94±0.30**
12	83.4±0.26$^{\#}$

Values are mean ± SEM. $p<0.05$, n=3

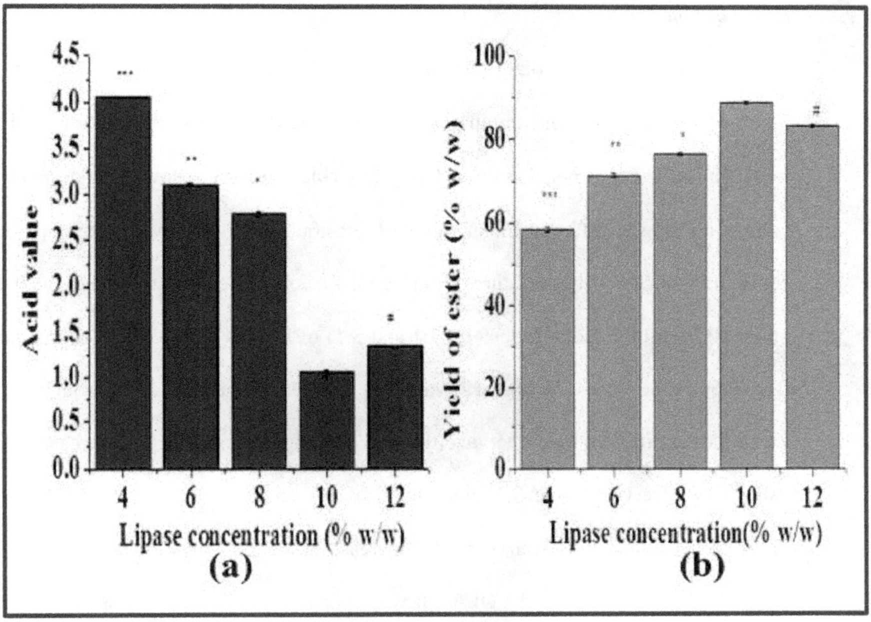

Figure 17.Effect of lipase (NS 435) concentration (4% to 12% w/w) (a) on acid value of the reaction of FA and COFA at 3:1 molar ratio and at 60⁰ C temperature at a constant time (5h) (b) on the yield of ester of the reaction of FA and COFA at 3:1 molar ratio and at 60⁰C temperature at a constant time (5h)

(a) Values are mean ± SEM, $p<0.05$, n=3).

(b) Values are mean ± SEM, $p<0.05$, n=3).

Statistical analysis:

#Significant at <0.05 in lipase concentration 10% vs 12% w/w

*Significant at < 0.05 in lipase concentration 10% vs 8% w/w

**Significant at <0.05 in lipase concentration 10% vs 6% w/w

***Significant at <0.05 in lipase concentration 10% vs 4% w/w

Effect of reaction time:

Reaction time played a pivotal role in the yield of the reaction product. In the present study, it was observed that the maximum acid value decreased as well as the highest yield of ester was obtained at 5 h of reaction time. **Table 29 (videFigure.18)** showed the decrease in acid value with time at different molar ratios of substrate with 10% (w/w) immobilized lipase concentration at 60°C temperature. From figure 4 a noticeable decrease in the acid value was observed for all the molar ratios (1:1,3:1 and 5:1) of FA and COFA at 5 h of reaction time. But when in the Figure 18, the acid values of different molar ratios were compared among each other at different time intervals (0 h to 6 h), a significant ($p<0.05$) decrease of the acid value was observed at the 3:1 molar ratio of FA and COFA. At 6 h of reaction time, figure.18 showed a significant ($p<0.05$) increase in the acid value for all the molar ratios (1:1, 3:1 and 5:1) of FA and COFA. It was evident from the figure that in case of all the ratios the optimum time of reaction was 5 h. The hydroxyl group present in the ricinoleic acid, a major fatty acid present in COFA, renders a higher viscosity and lesser reactivity to COFA. Therefore a higher reaction temperature and a longer reaction time were required to obtain same yield of ester.

Table 29.Changes in acid value at different molar ratios (1:1,3:1 and 5:1) of FA and COFA using NS 435 as biocatalyst (10%w/w) at different time intervals (0 h to 6 h) at 60°C

Time(h) ↓	Acid Value		
	Molar Ratio(FA:COFA)		
	1:1	3:1	5:1
0	87.20±2.85	87.96±2.40	86.89±1.07
1	31.31±0.88	16.32±1.43	20.34±0.50
2	17.13±0.32	11.46±1.14	14.2±0.47
3	14.57±0.56	9.26±0.67	12.23±0.28
4	9.86±0.32	5.09±0.70	7.78±0.07
5	8.91±0.05	**1.37±0.36**[*]	4.52±0.50
6	10.60±0.30	5.65±0.56	8.06±0.20

Values are represented as mean ± SEM. $p<0.05$,.n=3

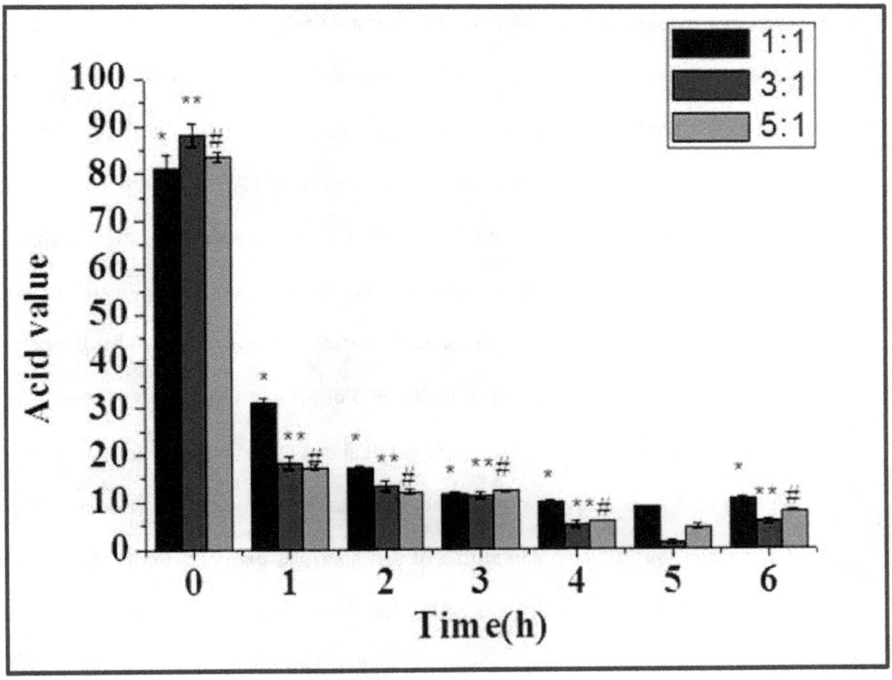

Figure 18.Changes in acid value with different time intervals (0h to 6h) at different molar ratios of FA and COFA (1:1, 3:1 and 5:1) at 60°C temperature and at 10% (w/w) lipase concentration. Values are mean ± SEM, n=3. Superscripts in the different bars indicate significant differences in results ($p < 0.05$)

Characterization of ester of COFA corresponding to the pure COFA

COFA used for the present study contained 88.80%w/w ricinoleic acid. Analysis of fatty acid composition of ester showed that almost 88.68% w/w of the ricinoleic acid of pure COFA were esterified with FA. The fatty acid profiles of the pure COFA and the ester of FA and COFA produced at different reaction parameters were shown in the **Tables 30,31 and 32**.From the Table 30, the fatty acid profile of FA-COFA esters produced from different molar ratios (1:1,3:1 and 5:1) of FA and COFA were determined at different time intervals (0 to 5h). Comparative study, as shown in the Table 30, revealed a small variation in the results of the proportion of ricinoleic acid. Moreover it was observed that the best yield of the reaction was obtained at 3:1 of molar ratio of FA and COFA at 60°C temperature with 10% (w/w) of immobilized lipase concentration and 5 h of reaction time where the highest amount of ricinoleic acid was esterified. Because of the lower yield of ester at 6 h reaction temperature, further fatty acid profile of that hour was not done. Furthermore, there was a trend in the initial hours of the reaction was found to be observed through the fatty acid composition of the ester that the palmitic acid content was high in comparison to ricinoleic acid and by the time the content of the palmitic acid was found to be decreased gradually. Subsequently, the ricinoleic acid content was higher with increasing reaction time.This kind of phenomena might be caused due to the high reactivity higher chain length of the ricinoleic acid in comparison to palmitic acid. Hence the reactivity of the palmitic acid is higher than ricinoleic acid.

Table 30.Fatty acid composition of castor oil fatty acid (COFA) and FA-COFA esters obtained at 60^{0}C, using different molar ratios of FA and COFA, at different reaction times

Molar Ratio of (FA:COFA)	Time of reaction (h)	Fatty acid composition of pure COFA and FA-COFA ester (%w/w)				
		16:0	18:0	18:1	18:2	Ricinoleic acid
COFA	-	1.99±0.01	1.14 ±0.01	3.68 ±0.12	4.74 ±0.07	88.80±0.16
1:1	1	30.42±0.31	1.58 ±0.01	18.24±0.15	9.34 ±0.12	32.93 ±0.37
	2	14.63±0.69	1.60±0.03	8.53 ±0.16	6.27 ±0.17	66.25 ±0.42
	3	10.40±0.04	1.54 ±0.04	7.84±0.14	2.36 ±0.08	71.74±0.88
	4	6.29±0.04	2.03±0.03	3.72 ±0.04	1.45 ±0.01	84.82 ±0.16
	5	5.11 ±0.01	1.89 ±0.01	3.78 ±0.05	1.39 ±0.02	86.29 ±0.95[a]
3:1	1	35.75±0.33	1.71 ±0.03	18.76±0.06	10.52±0.16	35.19 ±0.44
	2	17.73±0.13	1.58± 0.01	8.71 ±0.05	6.56 ±0.06	66.16 ±0.56
	3	11.87±0.01	2.16 ±0.01	7.79 ±0.03	2.49 ±0.03	75.88 ±0.10
	4	6.57 ±0.04	1.72 ±0.01	3.86 ±0.03	1.69 ±0.01	85.91 ±0.34
	5	5.14 ±0.03	1.12±0.01	3.25 ±0.10	1.98 ±0.01	**88.68±0.03**
5:1	1	38.43±0.19	1.77 ±0.01	18.68±0.01	11.54±0.02	30.07 ±0.02
	2	19.43±0.06	1.70 ±0.02	8.95 ±0.07	10.14±0.03	59.13±0.06
	3	13.12±0.04	2.37 ±0.04	7.84 ±0.03	7.13 ±0.03	65.72±0.04
	4	7.65±0.05	1.82 ±0.03	3.45 ±0.06	6.87 ±0.05	70.41 ±0.34
	5	9.80 ±0.05	2.24 ±0.05	4.62±0.05	6.82 ±0.04	76.47 ±0.05[b]

Values are represented as mean ± SEM .n=3

[a]Significant at<0.05 in all the groups at 5h reaction time at the molar ratio of 3:1 vs 1:1

[b]Significant at<0.05 in all the groups at 5h reaction time at the molar ratio of 3:1 vs 5:1

*FA and COFA- FA-Furfuryl alcohol, COFA-Castor oil fatty acid

From the **Table 31**, the effect of different temperature(45°C to 65°C) on fatty acid profile of the FA-COFA ester were determined by keeping the molar ratio (3:1), immobilized lipase concentration (10% w/w) , and the reaction time constant (5h). The comparative study showed (Table 31) significant ($p<0.05$) higher yields of ester (88.68% w/w) on the basis of ricinoleic acid at 60°C temperature. The yields of ester for the rest of the temperatures (45°C, 50°C, 55°C and 65°C) were relatively lower than 60°C. Generally enzymes found in solvent free environments become more stable, as the temperature rises up to a limit .In most of the cases, higher temperatures affect the lipase activity and it could be decreased significantly due to its denaturation. In addition, a lipase-catalyzed esterification reaction rate could be decreased gradually with increasing temperature, probably due to the effect of a synergic inhibition between temperature and alcohol. According to a report, lipase catalyzed esterification reactions are usually optimized in 60°C temperature [144].Hence the optimized reaction temperature for the present study is 60°C.

Table 31.Effect of temperature on the fatty acid composition of ester at a particular molar ratio of FA: COFA* (3:1) at a constant time(5h) at 10%(w/w) lipase concentration

Temperature (°C)	Fatty acid composition FA-COFA ester (%w/w)				
	16:0	18:0	18:1	18:2	Ricinoleic acid
45	17.84±0.1	1.60±0.02	8.77±0.02	6.48±0.01	65.60±0.01[a]
50	11.86±0.01	2.19±0.02	7.81±0.03	2.44±0.01	75.76±0.01[b]
55	10.14±0.05	1.59±0.02	6.52±0.04	5.96±0.02	75.97±0.02[c]
60	5.14 ±0.03	1.12±0.01	3.25 ±0.10	1.98 ±0.01	**88.68±0.03**
65	29.37±0.08	1.69±0.01	18.48±0.01	10.15±0.01	59.47±0.02[d]

Values are represented as mean ± SEM. n=3

[a]Significant at<0.05 in temperature 60°C vs 45°C

[b]Significant at<0.05 in temperature 60°C vs 50°C

[c]Significant at<0.05 in temperature 60°C vs 55°C

[d]Significant at<0.05 in temperature 60°C vs 65°C

*FA: COFA- FA-Furfuryl alcohol, COFA-Castor oil fatty acid

Table 32,revealed the effect of immobilized lipase concentration (4% to 12% w/w) on the fatty acid composition of FA-COFA ester where the parameters such as the reaction temperature (60^{0}C) and the molar ratio (3:1) of FA and COFA were kept constant because the lipase performs best at 60^{0}C.The comparative study showed higher yield (88.68 %w/w) of ester of FA-COFA siginificantly ($p<0.05$), on the basis of ricinoleic acid at 10%(w/w) of lipase concentration. Rest of the immobilized lipase concentrations (4%, 6%, 8% and 12% w/w) showed (Table 32)relatively lower yield of ester of FA-COFA in comparison to 10% (w/w) immobilized lipase concentration. This phenomenon might have happened due to the higher enzyme loading (12%w/w) which could affect the reaction rate as well as the reaction yield.

Table 32.Effect of lipase concentration (%w/w) on the fatty acid composition of ester at a particular molar ratio of FA: COFA (3:1) at a constant time (5h) and temperature (60^{0}C)

Lipase concentration (%w/w)	Fatty acid composition of FA-COFA* ester (%w/w)				
	16:0	18:0	18:1	18:2	Ricinoleic acid
4	18.5±0.30	2.67±0.05	15.06±0.01	3.36±0.02	60.89±0.01[a]
6	10.21±0.11	1.17±0.01	5.88±0.03	1.35±0.02	71.72±0.04[b]
8	9.71±0.01	1.18±0.03	4.89±0.05	1.32±0.03	76.72±0.05[c]
10	5.27±0.11	1.19±0.09	3.23±0.13	2.1±0.02	**88.68±0.10**
12	7.76±0.41	1.16±0.18	3.53±0.08	2.16±0.17	83.68±0.04[d]

Values are represented as mean ±SEM. n=3

[a]Significant at<0.05 in lipase concentration at 10% vs 4%(w/w)

[b]Significant at<0.05 in lipase concentration at 10% vs 6%(w/w)

[c]Significant at<0.05 in lipase concentration at 10% vs 8%(w/w)

[d]Significant at<0.05 in lipase concentration at 10% vs 8%(w/w)

FTIR analysis of COFA,FA and FA-COFA esters

FTIR spectra of COFA, FA and FA-COFA esters were compared to confirm the formation of the ester. As a result, the FTIR spectrum of COFA showed C=O stretching band at 1714 cm^{-1} and the presence of a hydroxyl group in the COFA structure was shown 3392.4 cm^{-1} wave number where else the reaction product, a sharp and long carbonyl C=O ester transmittance peak was observed at 1734 cm^{-1}(**Table 33 vide Figure 19 a, b and c**). The presence of CH_2bending at 743 cm^{-1} wave number indicates long-chain compound. The, C=C transmittance peak was clearly present because it, showed a weak transmittance peak due to only a small amount of alkene was present[110]. Other absorbance indicates C–H alkane bond and bend alkane at transmittance peaks 2926 cm^{-1}, 2854 cm^{-1} and 1464 cm^{-1}, respectively. There were several resemblances between the peaks of FA and ester of COFA and FA which confirmed the ester formation.

Table 33.FTIR peak assignment for COFA, FA and FA-COFA ester

Group of class	**Range(cm^{-1})**		
	COFA	**FA**	**ESTER**
Carbonyl (C=O)	1714	-	1734
Hydroxyl (-OH)	3391.4	3351.1	3393.8
Aromatic	-	1505.6	1503.9
Alkane (-CH_2-)	1463.7	-	1459.1
Alkene (C=C)	1410.7		-
O-H bending and C-O stretching vibrations of tertiary alcohol	-	1417.0	-
	-	1221.3	1221.1
		1148.8	1150.2
		1076.9	1078.7
		1010.3	1012.1
C-H bending	-	914.3	916.6
	885	884.9	885.0
	724	744.9	743.8

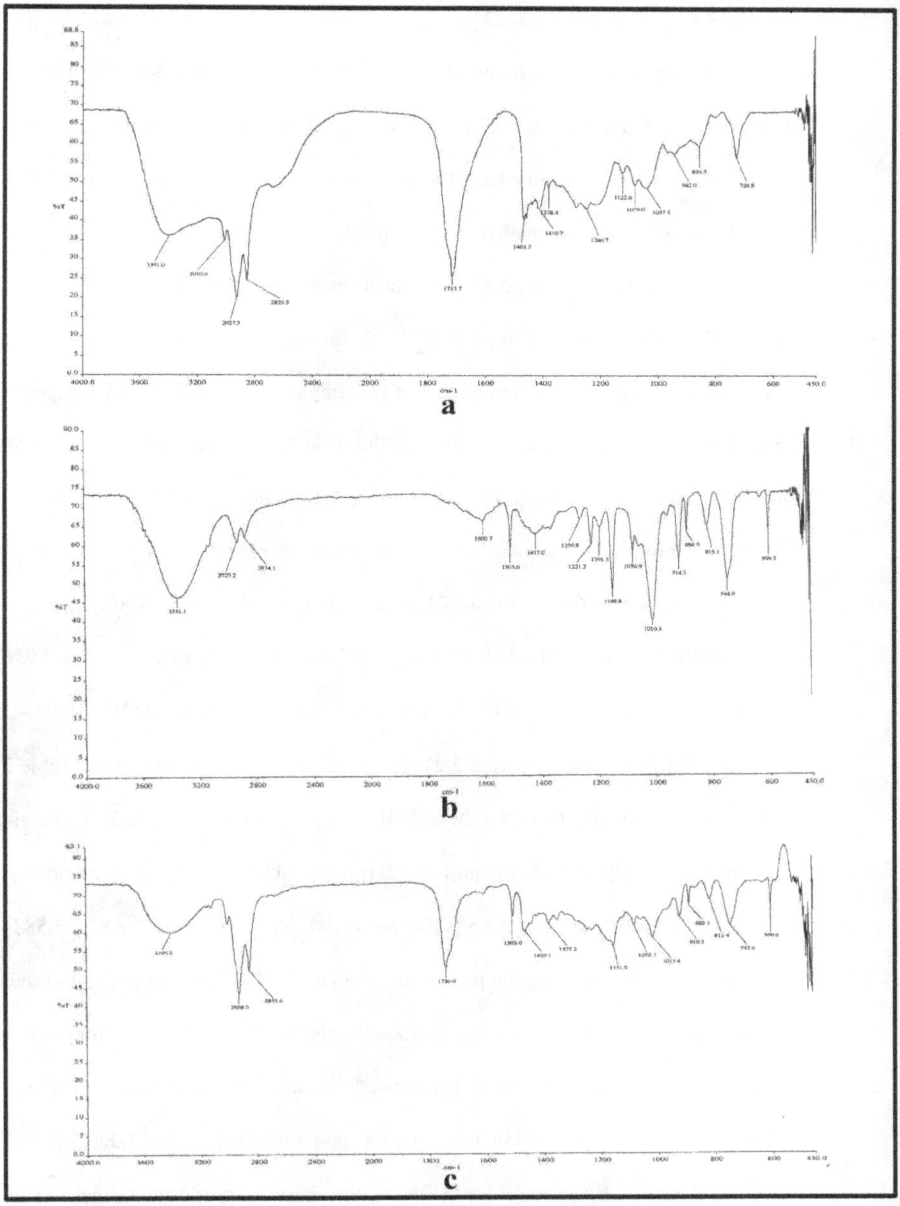

Figure 19. FTIR peak assignments for (a)COFA;(b) FA and (c) FA-COFA ester

^{1}H NMR and ^{13}C NMR

Table 34(vide Figure 20 a, b and c) precisely described the ^{1}H NMR spectra of COFA, FA and ester of FA and COFA with peak assignments. The signal ranging from 3.641ppm to 3.662 ppm in the spectrum of COFA as shown in Table 34 [vide Figure 20 a]was the distinct property of the methylene proton attached to the hydroxyl (C**H**-OH),while the signals at 2.18ppm and 1.4 ppm were related to the methylene protons of C-11 and C-13 (**CH_2**.CHOH-**CH_2**) , respectively [142].The signal ranging from 0.88ppm to 0.93ppm in the spectrum of COFA, as shown in table 4, was methylene proton[-CH_3(**3H**)]. Signals ranging from 1.308ppm to 1.554ppm, as shown in the table 4, were the signals of total **18H** which was attached at-CH_2- group of COFA. Signal at 1.626 ppm of COFA, as shown in the Table 32, was the proton (**1H**) attached to the –OH group of this fatty acid. In case of COFA, the signal ranging from 1.653ppm to 1.674 ppm, as shown in the Table 34, were the two protons (**2H**) attached to the -CH_2- β acid group. Signals of COFA ranging from 2.056ppm to 2.077ppm, as shown in the Table 34, were the protons (**2H**) attached to the allylic -CH_2-.The signals of the COFA ranging from 2.211ppm to 2.288 ppm, as shown in the Table 34, were the protons (**2H**) attached to other allylic -CH_2-.Signals of the COFA, ranging from 2.340ppm to 2.390ppm were the protons (**2H**) attached to -CH_2-α to acid group. Rest of the signals of the COFA ranging from 5.396ppm to 5.436 ppm and 5.562ppm to 5.588ppm, as shown in the Table 34, were the protons attached to the vinylic H .The signal of the olefinic protons at about 5.5 ppm was splitted following the proximity of the hydroxyl group to the double bond.^{1}H NMR spectrum of FA showed the signals of protons at $\delta = 7.42$ ppm (**5H**), 6.36 ppm (**4H**), 6.32 ppm (**3H**), 1.76 ppm-OH and 4.40 ppm -CH_2- (Table 33).^{1}H NMR of ester of FA and COFA (Table 34vide Figure 20c) showed significant changes in comparison to the pure COFA (Table 34).At the signals of FA-COFA ester ranging from 0.79ppm to

0.81 ppm (Table 34 vide Figure 20 c), there was a chemical shift in comparison to pure COFA. Signals of FA-COFA ester ranging from 1.185ppm to 1.393 ppm, as shown in the Table 34(vide Figure 20 c), were the protons associated with nine -CH_2 - (**18H**).Signal at 1.547 ppm , as shown in the Table 34, was the proton for –OH group (**1H**) which was the special characteristics of FA-COFA ester. The same signal was for the proton associated with -CH_2- β of acid group (**2H**).In case of the ester, proton associated with allylic -CH_2- (**2H**) was observed at the signal of 1.965 ppm and the protons with the other allylic -CH_2- were seen at the signal ranging from 2.117ppm to 2.158 ppm (Table 34).The signal ranging from 2.230ppm to 2.280 ppm of FA-COFA ester were the protons attached to the -CH_2 - α group of acid (Table 34).The signal of FA-COFA ester ranging from 2.117ppm to 2.158ppm , as shown in the table 34, was the proton for –CH attached to the –OH group. The ester showing the signal (Table 34) at 5.312ppm and 5.470ppm were the vinylic protons. The signals at 4.990 ppm in the spectrum of ester were the protons (**2H**) for -CH_2- group of furfuryl ring and rest of all signals (6.300ppm, 6.327ppm and 7.354ppm) were for the protonsof 3H of furfuryl ring which were incorporated into the COFA(Table 34). Hence the ester formation was confirmed by ^{1}H NMR spectral analysis [142].

Table 34.^{1}HNMR peak assignment for COFA, FA and FA-COFA ester

Assignment	(δ) ppm		
	COFA	FA	FA-COFA Ester
$-CH_3$(3H)	0.88-0.93	-	0.792-0.814
$-CH_2$-(18-H)	1.308-1.5549	-	1.185-1.393
-OH	1.626	-	1.547
-CH2-β to acid group	1.653-1.674	-	1.547
Allylic-CH_2-	2.056-2.077	-	1.965
Allylic -CH_2-	2.211-2.288	-	2.117-2.158
-CH_2 - α to acid group	2.340-2.390	-	2.230-2.280
-CH attached to –OH	3.641-3.662	-	2.117-2.158
Vinylic –H	5.396-5.436	-	5.312-5.349
Vinylic –H	5.562-5.588	-	5.470
-OH	-	1.760	
-CH_2-	-	4.637	4.990
-H	-	6.323	6.327
-H	-	6.365	6.330
$CHCl_3$ residual peak	-	7.283	-
-H	-	7.425	7.354

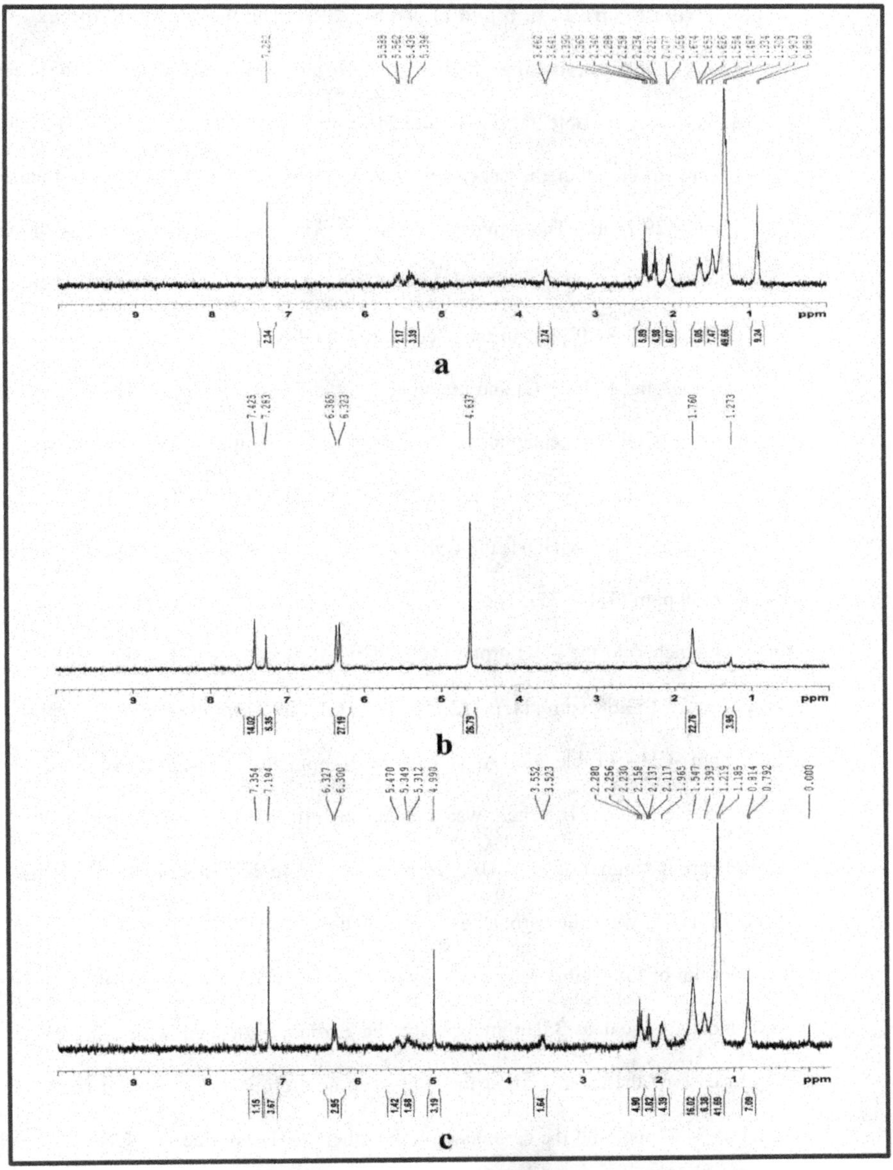

Figure 20.^{1}HNMR peak assignment for (a) COFA;(b) FA and (c) FA-COFA ester

Table 35(vide Figure 21 a, b and c) precisely described the ^{13}C NMR spectra of COFA, FA and FA-COFA ester with peak assignments. The signal at 13.70ppm in the spectrum of COFA, as shown in Table 35(vide Figure 21 a), was the carbon for $-CH_3$ **(1C)**.Carbon for 5 $-CH_2-$ were observed in the spectrum of COFA (Table 35) at the signals ranging from 28.55ppm to 29.21ppm.The signals at 22.24ppm, 31.38ppm and 25.31ppm, as shown in the Table 34, were the carbons of COFA for other two $-CH_2-$ in the aliphatic chain i.e, total **3C**.The signals at 34.9 ppm in the spectrum of COFA, as shown in the Table 35, was the carbon for allylic $-CH_2-$ **(1C)** and the signal at 26.95ppm was the carbon for other allylic $-CH_2$ group of COFA. The signals at 33.62 ppm and 24.29ppm in the spectrum of COFA, as shown in the Table 35, were carbons for $-CH_2-$ α**(1C)**and $-CH_2$-β **(1C)** to acid group respectively. Besides –CHOH, the Carbon for the other $-CH_2-$ groups was observed at the signal 36.39 ppm (Table 35). The signal at 71.27 ppm, as shown in Table 35, was the carbon for –CH attached to the –OH group **(1C)**. ^{13}C NMR spectrum of FA showed the signals of carbons at δ= 153.68 ppm **(1C)**, 142.18 ppm **(1C)**, 109.91ppm **(1C)**, 107.36 ppm **(1C)** and 57.00 ppm $-CH_2-$ (Table 35). At signals 14.01 ppm in the spectrum of FA-COFA ester (Table 35vide Figure 21c), there was a chemical shift in comparison to pure COFA in case of $-CH_3$ group. Signal of FA-COFA ester ranging from 22.56ppm to 29.50 ppm ,as shown in the Table 35, were the carbons associated with nine $-CH_2-$ **(9C)**. Signal at 31.78 ppm in the spectrum of FA-COFA ester , as shown in the Table 35 ,was the carbon for $-CH_2-$ group **(1C)** and the signal at 35.3ppm was for the carbon associated with other allylic $-CH_2-$ **(1C)**.In case of the ester, carbon for $-CH_2$-α to acid group was observed (Table 35)at the signal of 34.08ppm and the carbons with the other $-CH_2-$ beside –CHOH were seen at the signal 36.8 ppm (Table 35).The signal at 71.49 ppm in the spectrum of FA-COFA ester was the carbon for -CH attached to –OH ,which is the special characteristic of the ester (Table

35).The signal at 133.39 ppm and 125.16 ppm ,as shown in the Table 35,were the vinylic carbon. The ester showing the signal (Table 35) at 57.83 ppm was the carbon between ester -O- and the furfuryl ring (**1C**). The signal at 173.41 ppm in the spectrum of ester was the ester carbon (**1C**) .Rest of the signals for **1C**i.e, total (**4C**) of furfuryl ring (107.67ppm,110.3 ppm,149.62ppm and 159.99 ppm) were incorporated into the COFA(Table 35).Therefore, FA-COFA ester formation was also confirmed by the ^{13}C NMR spectral analysis.

Table 35.^{13}C NMR spectrum assignment for COFA, FA and FA-COFA ester

Assignment	**(δ) ppm**		
	COFA	**FA**	**FA-COFA Ester**
$-CH_3$ (1C)	13.70	-	14.01
5-CH_2-(5C)	28.55-29.21	-	22.56-29.50
2-CH_2-(3C)	22.24	-	31.78
2-CH_2-(3C)	31.38	-	35.3
2-CH_2-(3C)	25.31	-	34.08
Allylic -CH_2- (1C)	34.9	-	36.8
Allylic -CH_2- (1C)	26.95	-	71.49
-CH_2-α to acid group (1C)	33.62	-	-
-CH_2 - β to acid group (1C)	24.29	-	-
-CHOH (1C)	36.39	-	-
-CH attached to the	71.27	-	
Vinylic carbon (1C)	133.29	-	133.29
Vinylic carbon(1C)	125.16	-	125.16
Ester carbon (1C)	173.41	-	173.41
Carbon attached to -OH group of FA(1C)	-	57.00	57.83
1C	-	107.36	107.67
1C	-	109.91	110.3
1C	-	142.18	149.62
1C	-	153.68	153.99

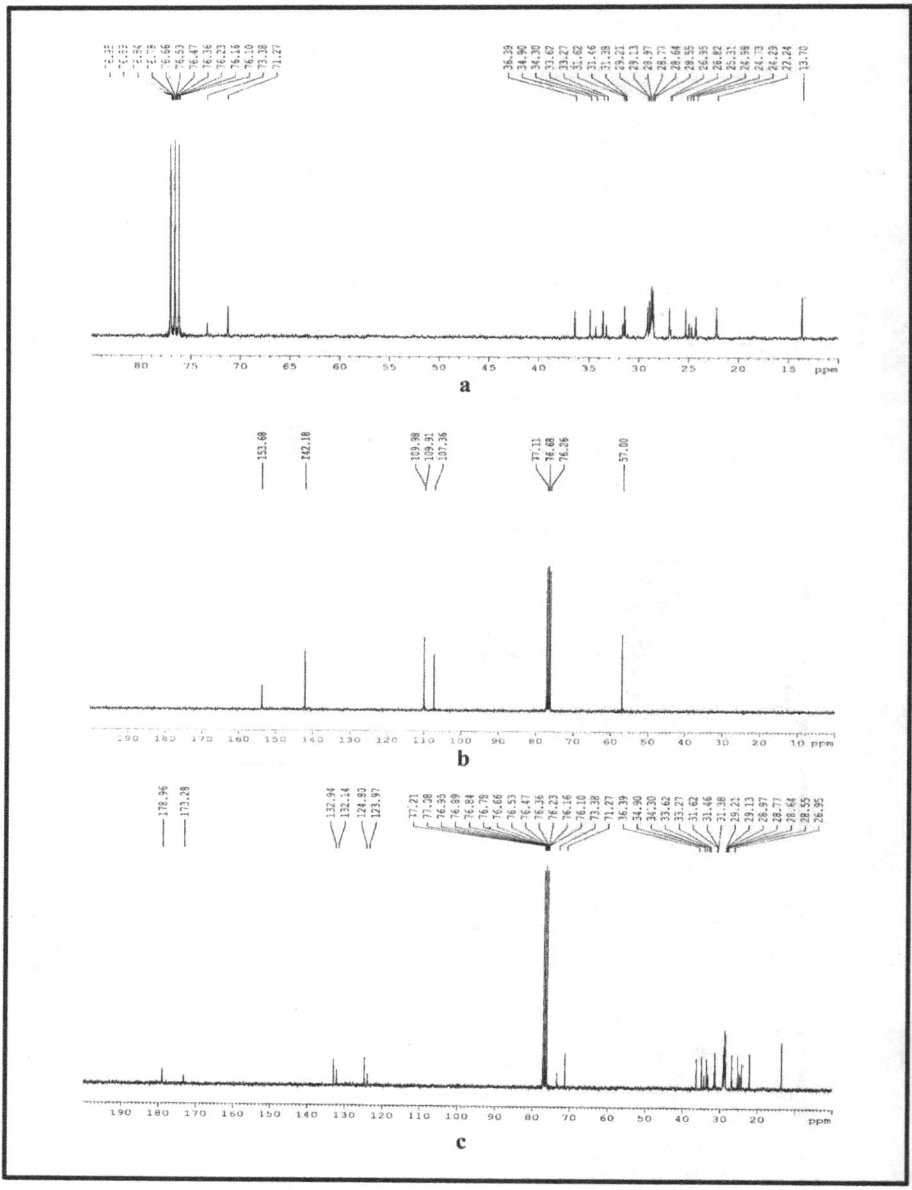

Figure 21.^{13}C NMR spectrum assignment for (a)COFA;(b) FA and (c) FA-COFA ester

Physical characterization of COFA and ester of FA and COFA:

The physical characterization of the COFA and ester of COFA were listed in the **Table 36**. The physical properties of the COFA and ester of the FA and COFA obtained were comparable with the results obtained by other researchers. The flash point of ester of FA and COFA was 242 °C. The pour point value was less than -23°C. The viscosity of COFA is 246.8 cP, whereas for ester of FA and COFA, it was 17.9 at 40°C. The density of the COFA and ester of FA and COFA was almost similar.

Table 36.Physical characterization of COFA and ester of FA and COFA

Properties	Values	
	COFA	Ester
Flash Point (°C)	260±0.47	242±0.82
Pour Point (°C)	-21±0.81	-23±0.471
Viscosity(cP at 40 °C)	246.8±0.86	17.9±0.68
Density (g/cm^3) at 30°C	0.961±0.007	0.970±0.008

Values are represented as mean ±SEM. n=3

BIODEGRADABILITY STUDY OF FA-COFA ESTER

Many bacteria have been reported to degrade different type of lubricant oils and much work has also been carried out on the degradation of vegetable oil [143,145]. Studies on degradation of newly synthesized FA-COFA ester were carried out to check the biodegradability of the product.

It has been reported [146] that castor oil is biodegradable in nature and any product synthesized from this oil would be biodegradable. It would be more advantageous if the normal soil flora degrades it. *Ochrobactrum pseudintermedium* showed very efficient degradation of lubricating oil and hence was selected for the degradation study of the FA-COFA ester.

FA-COFA ester ranging from 0.1g to 0.5g (e.g.0.1g of ester added to the 10ml of BH medium) was provided as the sole source of carbon for the growth of the microorganism. An increased growth of microorganism was observed in the spectrophotometric study where the absorbance was measured at 600nm using Shimadzu (1601A) UV-VIS Spectrophotometer. Loss of weight of the ester was also observed due to the biodegradation of the ester **(Table 37 vide Figure 22)**. Spectrophotometric study, as shown in the **Table 37 (vide Figure 22)**, revealed significantly ($p<0.05$) higher growth of microorganism at 3% (w/v) concentration of the ester at the 5th day of the study, whereas, the rest of the ester concentration i.e,1%,2%,4% and 5% (w/v) in the growth medium showed relatively lower growth of organism in the culture media. But in each concentration of the ester, ranging from1% to5% (w/v),as shown in the Figure 22,all the experimental set showed the significant ($p<0.05$) growth of microorganism against its control set. It was a noticeable part of the study that higher the concentration of the ester (e.g.4% and 5% w/v), the growth of the microorganism becomes static. Thus, the experiments proved the biodegradable nature of the ester. From the study it might also be inferred that higher concentrations of the product were not suitable for the growth of the organisms.

Table 37.Spectrophotometric study (Absorbance at 600 nm) for the growth of microorganism (*Ochrobactrum pseudintermedium*)in the BH media where FA-COFA ester was added at different concentrations ranging from 1% to 5% (w/v) and the study was continued for five days .Where C: control set and E: experimental for each concentration of FA-COFA.

Days	Absorbance at 600nm									
	Reaction Product(w/v)									
	1%		2%		3%		4%		5%	
	C	E	C	E	C	E	C	E	C	E
1stDay	0.02± 0.01	0.14± 0.01[a]	0.08± 0.01	0.24± 0.02[b]	0.11± 0.02	0.25± 0.02[c]	0.11± 0.01	0.21± 0.02[d]	0.12± 0.01	0.26± 0.01[e]
3rdDay	0.02± 0.01	0.52± 0.01[a]	0.08± 0.01	0.48± 0.01[b]	0.11± 0.02	0.25± 0.03[c]	0.11± 0.01	0.4± 0.02[d]	0.12± 0.01	0.68± 0.01[e]
5thDay	0.02± 0.01	0.49± 0.01[a]	0.08± 0.01	0.81± 0.06[b]	0.11± 0.02	0.84± 0.02[c]	0.11± 0.01	0.74± 0.01[d]	0.12± 0.01	0.71± 0.01[e]

Values are mean represented as ± SEM. P<0.05, n=3

[a]Significant at < 0.05 in 1st,3rd and 5th day C vs E in all group of 1%(w/w) concentration of ester.

[b]Significant at < 0.05 in 1st,3rd and 5th day C vs E in all groups 2%(w/w) concentration of ester.

[c]Significant at < 0.05 in 1st,3rd and 5th day C vs E in all groups3% (w/w) concentration of ester.

[d]Significant at< 0.05 in 1st,3rd and 5th day C vs E in all groups 4% (w/w) concentration of ester.

[e]Significant at<0.05 in 1st,3rd and 5th day C vsEin all groups 5% (w/w) concentration of ester.

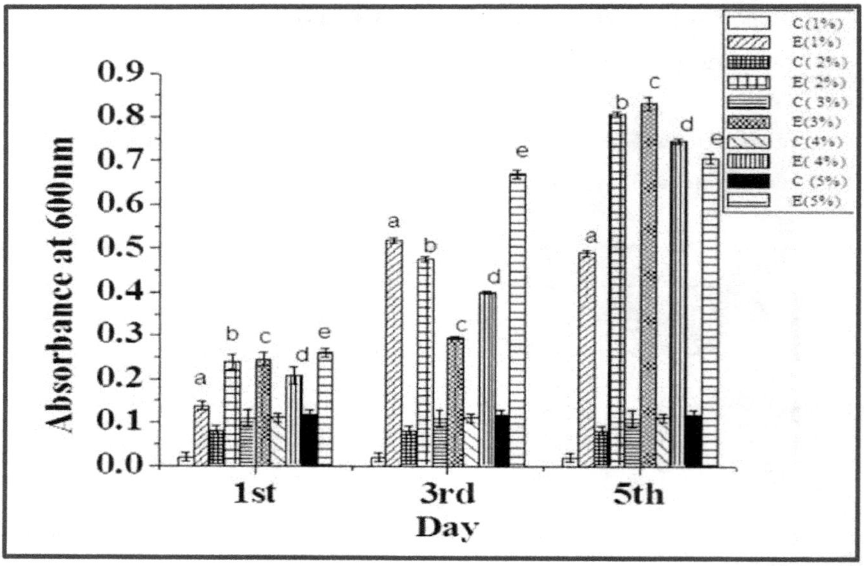

Figure 22.Spectrophotometric study (Absorbance at 600 nm) for the growth of microorganism (*Ochrobactrum pseudintermedium*)in the BH media where FA-COFA ester was added at different concentrations ranging from 1% to 5% (w/v) and the study was continued for five days .Where C: control set and E: experimental for each concentration of FA-COFA. Values are mean ± SEM. $p<0.05$, n=3

[a]Significant at < 0.05 in 1st,3rd and 5th day C vs E in all group of 1%(w/w) concentration of ester.

[b]Significant at < 0.05 in 1st,3rd and 5th day C vs E in all groups 2%(w/w) concentration of ester.

[c]Significant at < 0.05 in 1st,3rd and 5th day C vs E in all groups3% (w/w) concentration of ester.

[d]Significant at< 0.05 in 1st,3rd and 5th day C vs E in all groups 4% (w/w) concentration of ester.

[e]Significant at<0.05 in 1st,3rd and 5th day C vs E in all groups 5% (w/w) concentration of ester.

After 5th day, the loss of weight of the ester of biodegradation was determined from the gravimetric study, presented in **Table 38 (vide Figure 23)**. Figure 23showed significant amount ($p<0.05$) of weight loss at the 3% (w/v) concentration of the ester in comparison to its control set (3% w/v) .Moreover, greater loss of weight was found at 3% (w/v) concentration in comparison to all the concentrations (1%,2%,4% and 5% w/v) of the ester. In case of 1%, 2% and 3% (w/v) concentrations of ester, as shown in the figure 6, there was a noticeable amount of weight loss of ester whereas, in case of 4% and 5%, not much of loss of weight was found. Least amount of weight loss of the ester was found in the rest of the concentrations of the reaction product. Hence the optimized concentration of the reaction product for the biodegradation study was at 3% (w/v) concentration.

Since the studies were carried out in the mineral salt medium (BH medium) providing FA-COFA ester as the sole carbon source, the increased growth of bacteria in the media as well as decreasing weight of the reaction product proved the bio-degradability of the FA-COFA ester against *Ochrobactrum pseudintermedium.* The degradation of the product proved that the prepared ester was completely biodegradable which is extremely desirable in terms of environmental safety. In fact, greater degradation could be achieved with lesser incubation time, depending on the nature of the product prepared.

Table 38.Gravimetric study of the FA-COFA ester at different concentrations (1%, 2%,3%,4% and 5% w/v) after the five days study of microbial biodegradation. Control= weight (g) of the ester in the media without bacterial inoculums. Experimental= loss of weight (g) of the ester in the media containing bacterial inoculums

Reaction Product(w/v)	Control	Experimental
1%	0.11±0.01	0.06±0.06
2%	0.23±0.01	0.11±0.05
3%	**0.32±0.02[a]**	**0.15±0.04[b]**
4%	0.42±0.04	0.34±0.01
5%	0.51±0.08	0.46±0.05

Values are represented as mean ± SEM, n=3. $p<0.05$,

[a]Significant at < 0.05 in FA-COFA ester at 3% (w/v) concentration with the experimental group.

[b]Significant at< 0.05 in FA-COFA ester at 3% (w/v) concentration with the control group.

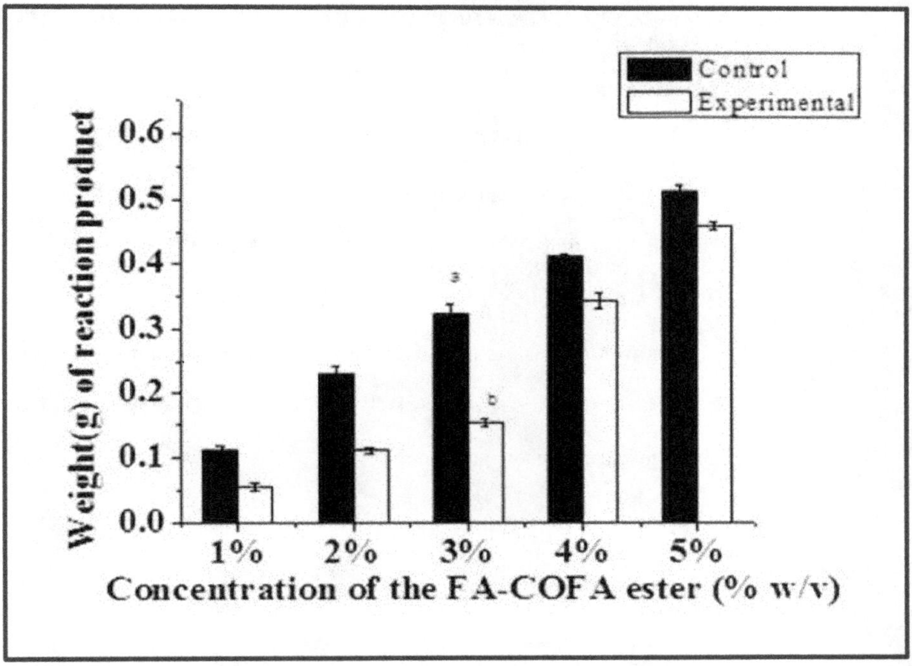

Figure 23.Gravimetric study of the FA-COFA ester at different concentrations (1%, 2%,3%,4% and 5% w/v) after the five days study of microbial biodegradation. Control= weight (g) of the ester in the media without bacterial inoculum. Experimental= loss of weight (g) of the ester in the media containing bacterial inoculum

Values are represented as mean ± SEM

[a]Significant at < 0.05 in FA-COFA ester at 3% (w/v) concentration with the experimental group.

[b]Significant at< 0.05 in FA-COFA ester at 3% (w/v) concentration with the control group.

PERFORMANCE EVALUATION OF THE FA-COFA ESTER PLASTICIZED PVC FILMS

X Ray diffraction analysis:

XRD analysis was performed to determine the crystallization, complexation and structure of the PVC-plasticizer matrix. **Figure 24** depicted the XRD diffractogram of pure PVC film and plasticized PVC films [PVC+(DBP: FA-COFA :: 8:2); PVC+(DBP:FA-COFA::6:4); PVC+(DBP :FA-COFA ::4:6); PVC+(DBP:FA-COFA::2:8)] respectively. Pure PVC film and PVC films with plasticizers (Figure 24) with all the concentrations [PVC+(DBP: FA-COFA :: 8:2); PVC+(DBP:FA-COFA::6:4); PVC+(DBP :FA-COFA ::4:6); PVC+(DBP:FA-COFA::2:8)]showed amorphous structure.

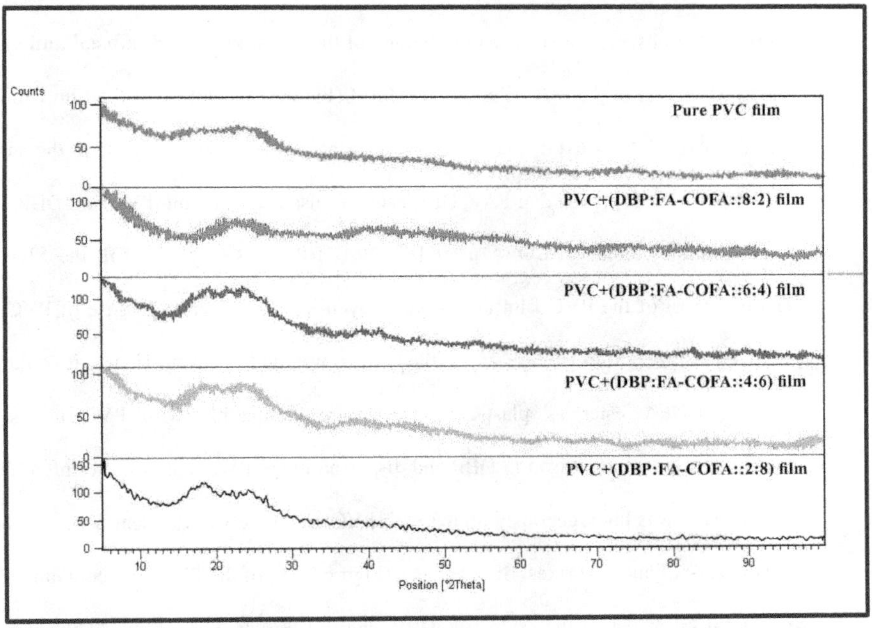

Figure 24.XRD analysis of Pure PVC film, PVC film containing FA-COFA ester at different concentration in combination with DBP plasticizer i.e. 20–80%w/w) [PVC+(DBP: FA-COFA :: 8:2) film ,PVC+(DBP:FA-COFA::6:4) film ,PVC+(DBP :FA-COFA ::4:6) film and PVC+ (DBP: FA-COFA :: 2:8) film]

Scanning electron microscopy (SEM):

The characteristics of the plasticized film composites strongly depend on the dispersion and compatibility of plasticizers in polymer matrix. Thus, the scanning electron microscopy was employed to observe the surface morphology of the films. **Figure 25 (a,b,c,d and e)** depicts the SEM images of the PVC film surface. SEM (Figure 25 d) image of the film with [PVC+ (DBP :FA-COFA ::4:6)] displayed an even surface which was analogous to the surface of pure PVC film. It proves that FA-COFA ester is miscible with both PVC and DBP. At 80% concentration of the FA-COFA ester [PVC+ (DBP: FA-COFA :: 2:8)], the SEM image (Figure 25 e) of the PVC film also showed even surface, whereas in case of PVC+(DBP: FA-COFA :: 8:2) film (Figure 25 b), the surface was quite uneven. Hence, it could be said that FA-COFA ester as plasticizer was more compatible with PVC at its higher concentration in comparison to DBP and dispersed in the PVC matrix uniformly [126]. FA-COFA ester was homogenously mixed with PVC and there was no clear boundary between plasticizer-polymer matrixes. SEM images (Figure 25 b) of the PVC film containing 20% of the FA-COFA ester and 80% of DBP showed rough surface, probably the higher concentration of DBP caused to the aggregation of plasticizer in PVC film, leading to the partial phase separation during the film formation[110,149].

When the PVC blend was cast into the mold, solvent evaporation took place and as a consequence the polymer chains come into close vicinity. Henceforth, in plasticized PVC films the functional groups or long chains of the plasticizers would have major influences on the formation of hydrogen bonds, di-pole-di-pole bonds and Van der Waals bond [110,149].

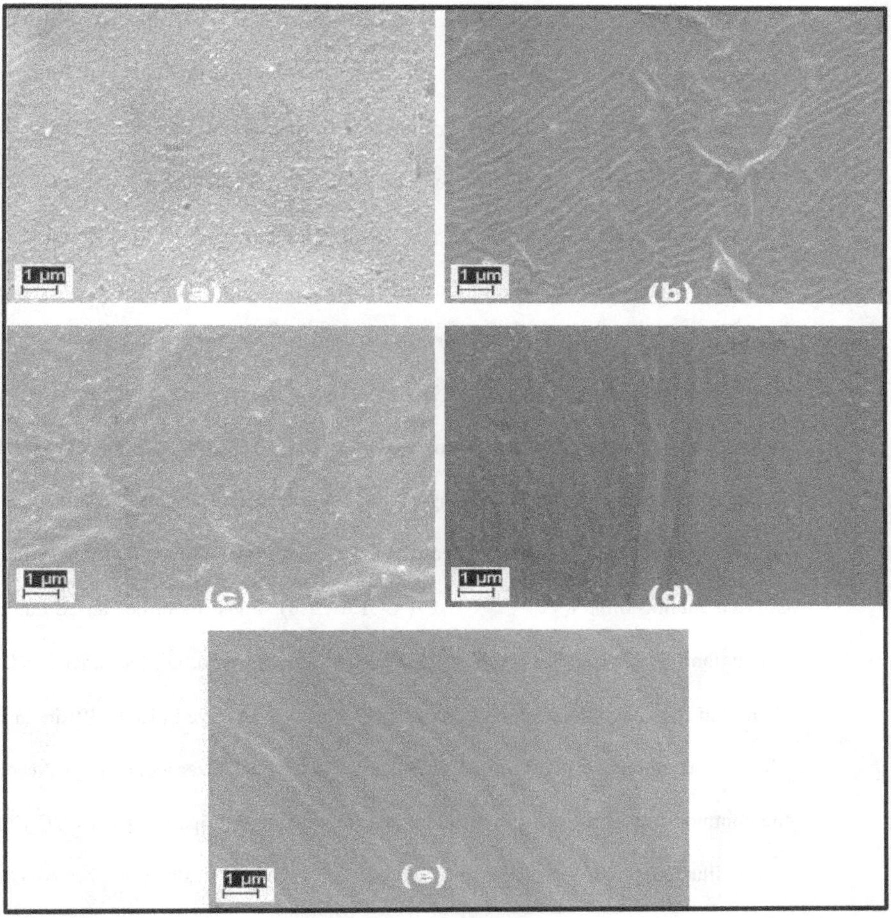

Figure 25.SEM images for surface morphology of (a) Pure PVC film, (b) PVC+(DBP: FA-COFA :: 8:2) film ,(c) PVC+(DBP:FA-COFA::6:4) film , (d) PVC+(DBP :FA-COFA ::4:6) film and (e) PVC+ (DBP: FA-COFA :: 2:8) film

Mechanical properties:

Mechanical properties of FA-COFA and DBP plasticized PVC films were evaluated by analyzing thickness, tensile strength, elongation at break and modulus elasticity. **Table 39** described the test results for the mechanical performances of PVC films. For each of the PVC films, there was a slight increase in the thickness with the increase of the concentration of plasticizers. According to the theory, tensile strength of PVC film would decrease if plasticizer content increased [126]. Lowest value of tensile strength was observed (Table 39) in case of the PVC film containing 80% of FA-COFA [PVC+ (DBP: FA-COFA ::2:8)] . The tensile strength of PVC was quite low even when the percentage of FA- COFA ester addition was in the range of 20-40% of total plasticizer (Table 39). Modulus properties were found to be decreased with increasing FA-COFA ester concentration but an increment of modulus property was observed in case of the PVC films containing 80% of the FA-COFA ester containing films PVC+ (DBP : FA-COFA ::2:8) which might be caused due to some interaction of the plasticizer with PVC at higher concentration. On the other hand, a huge change in the elongation property of the PVC film was observed (Table 39) in case of the PVC film containing 80% of the FA-COFA ester in the plasticizer blend [126]. According to the common rule, plasticizer could increase the elongation property of the PVC films. The results illustrated that the plasticizing efficacy of FA-COFA ester at higher concentration was better in comparison to the conventional plasticizer DBP in PVC film matrix [126]. The reason might be that PVC and FA-COFA interaction was better than PVC and DBP at higher concentration. Thus FA-COFA might be a good substitute for conventional phthalate plasticizers.

Table 39.Mechanical performance of PVC films

Film	Thickness (mm)	Tensile strength (MPa)	Elongation at break (%)	Modulus of elasticity (MPa)
Pure PVC film	0.69±0.03[a]	44.66±0.14[c]	35.03±0.01[e]	6.604±0.02[g]
PVC+(DBP:FA-COFA::8:2)	0.743±0.03	25.98±0.22	46.02±0.04	6.609±0.05
PVC+(DBP:FA-COFA::6:4)	0.75±0.03	17.09±0.1	70.55±0.01	6.509±0.01
PVC+(DBP:FA-COFA::4:6)	0.78±0.03	14.42±0.015	79.41±0.05	6.507±0.03
PVC+(DBP:FA-COFA::2:8)	**0.803±0.03[b]**	**9.96±0.23[d]**	**130.62±0.02[f]**	**10.605±0.01[h]**

Values are represented as mean ± SEM. n = 3

Superscripts in the different column indicate significant differences ($p < 0.05$)

Thermal Behavior:

Thermogravimetric curves of pure PVC film showed (**Figure 26 a**) a degradation between 200^{0}C and 300^{0}C corresponding to the loss of weight of the pure PVC film. Degradation of the pure PVC film started above 200^{0}C and a weight loss up to 40% was observed at 300^{0}C onwards. The TGA (Figure 26 a) curves of plasticized PVC films [PVC+(DBP: FA-COFA :: 8:2); PVC+(DBP: FA-COFA :: 6:4); PVC+(DBP :FA-COFA :: 4:6); PVC+(DBP : FA-COFA ::2:8)] showed almost same pattern in comparison to the pure PVC films. The substitution of the conventional plasticizer DBP with FA-COFA ester showed same kind of thermal stability from lower to higher concentration of FA-COFA ester concentration[128-130].Hence FA-COFA consists same kind of thermal stability as di-butyl phthalate as a plasticizer.

DTA (**Figure 26 b**) curves of plasticized PVC films [PVC+(DBP: FA-COFA :: 8:2); PVC+(DBP: FA-COFA :: 6:4); PVC+(DBP :FA-COFA :: 4:6); PVC+(DBP : FA-COFA ::2:8)] showed same glass transition temperature at 85°C as the pure PVC film . In the temperature range 200-300°C, a first-order endothermic curve can be observed which has been designated as the fusion, or "melting" point of the resin. PVC is partially syndiotactic, that's why the small crystals melt in the region of 200 to 300°C. DTA showing (Figure 26 b) the plateau at the starting of the curve relates to the de-chlorination of PVC with the formation and stoichiometric elimination of HCl. In addition there was an endothermic peak related to the polymer cross linking of C=C bond. Hence the PVC films containing increased percentage of FA-COFA ester showed almost same thermal stability in comparison to pure PVC films.

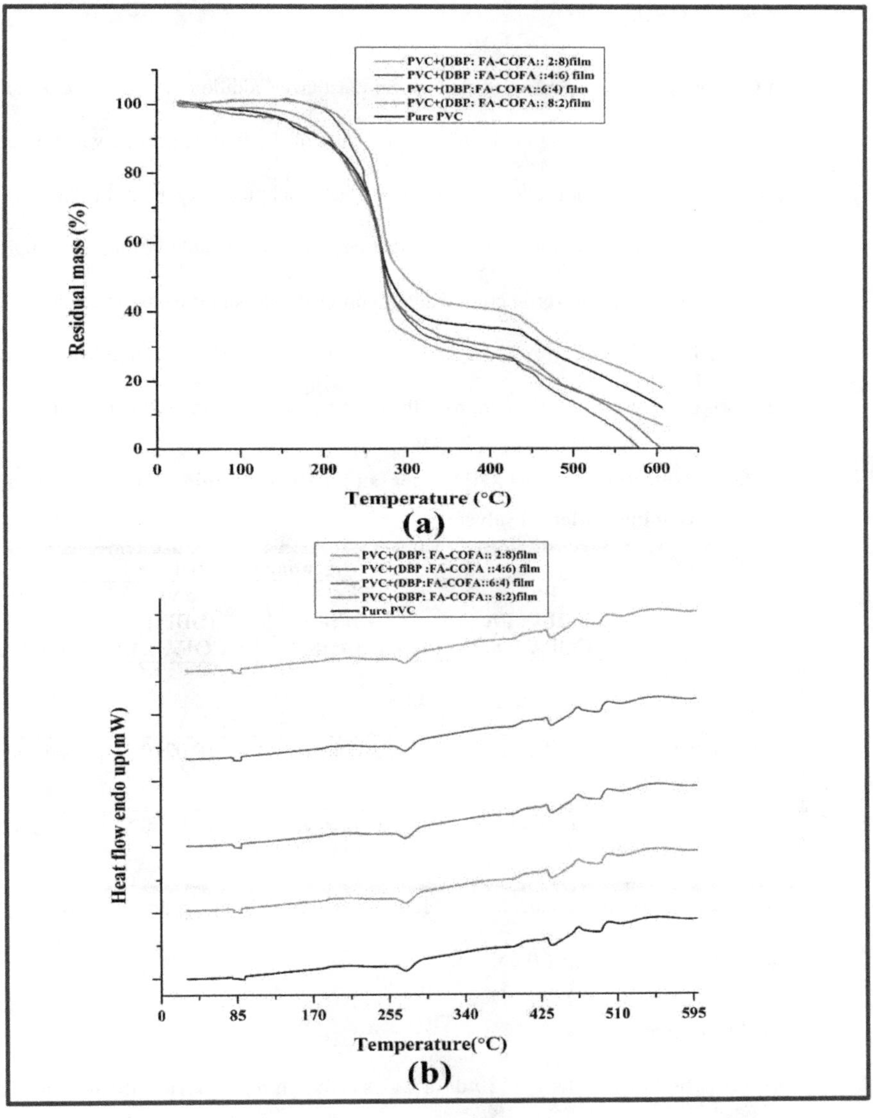

Figure 26.(a) Thermogravimetric analysis & (b) Differential thermal analysis of Pure PVC film, PVC+(DBP: FA-COFA :: 8:2) film, (PVC+(DBP:FA-COFA::6:4) film, PVC+(DBP :FA-COFA ::4:6) film and PVC+ (DBP: FA-COFA :: 2:8) film

Migration stability test:

PVC films with major portion of the natural plasticizer addition did not present (**Table 40**) water migration, indicating the hydrophobic nature of the film. However, when immersed in gasoline and petroleum ether, these plasticized samples presented (Table 40) a low migration resistance. It is very much interesting to observe (Table 40) that the migration of modified natural plasticizer in combination with DBP was significantly lower into these two solvents, indicating that FA-COFA ester incorporation at 80% in combination with 20% DBP was a successful effort to improve the plasticization property of the PVC matrix.

Table 40.Migration of FA-COFA ester and DBP containing PVC film of different concentration into different solvents

Solvent	Migration (% w/w)			
	PVC+ (DBP: FA-COFA :: 8:2)	PVC+ (DBP:FA-COFA::6:4)	PVC+ (DBP:FA-COFA ::4:6)	PVC+ (DBP :FA-COFA ::2:8)
Water	-	-	-	-
Petroleum ether	25.82±0.1[a]	23.02±0.3[a]	18.02±0.2[a]	**15.09±0.3[a]**
Gasoline	20.89±0.2[b]	16.08±0.2[b]	14.09±0.5[b]	**12.02±0.2[b]**

Values are represented as mean ± SEM. n = 3.Superscripts in the different column indicate significant differences ($p < 0.05$)

Soil Burial Degradation:

A short period of the study of 30 days was steered on the behavior of soil degradation of PVC films with plasticizer blend. Though short term studies are very much useful in portraying the soil degradation conditions (e.g., for 6 months or one year).The degradation increased with the increase in the FA-COFA content in the PVC films along with the conventional DBP plasticizer which is presented in the **Table 41 (vide Figure 27)** .The blend of PVC+(DBP : FA-COFA :: 2:8) showed a maximum degradation of 16.73% weight within a month whereas minimum degradation of only 1.11% was observed in case of PVC+(DBP: FA-COFA :: 8:2). Interestingly the PVC blend containing 40% of FA-COFA ester as plasticizer [PVC+ (DBP: FA-COFA:: 6:4)] also showed 12% degradation after 1 month. Soil microflora might be responsible for the degradation of these PVC blends as in the case of degradation of bioplasticizer based polymers[123]. The higher degradation nature of the PVC film could be attributed to the higher concentration of FA-COFA ester in the PVC films. In case of PVC+ (DBP: FA-COFA:: 8:2) degradation of only 1.11% clearly revealed that PVC and DBP was not degraded by soil microorganisms and only FA-COFA ester is biodegradable in the blend.

Table 41.Degradation of PVC+ (DBP: FA-COFA) blends in Soil.

Percentage of FA-COFA ester in PVC(%w/w)	**Loss of Weight(%w/w)**		
	10 Days	**20Days**	**30Days**
20	0.02±0.03	0.53±0.03	1.11±0.07
40	0.03±0.01	1.5±0.05	6.15±0.05
60	0.06±0.02	2.66±0.01	12.08±0.03
80	**0.08±0.01^{a}**	**6.02±0.25^{b}**	**16.73±0.12^{c}**

* Values are represented as mean ± SEM. n = 3.($p<0.05$)

Superscripts in the different column indicate significant differences ($p < 0.05$)

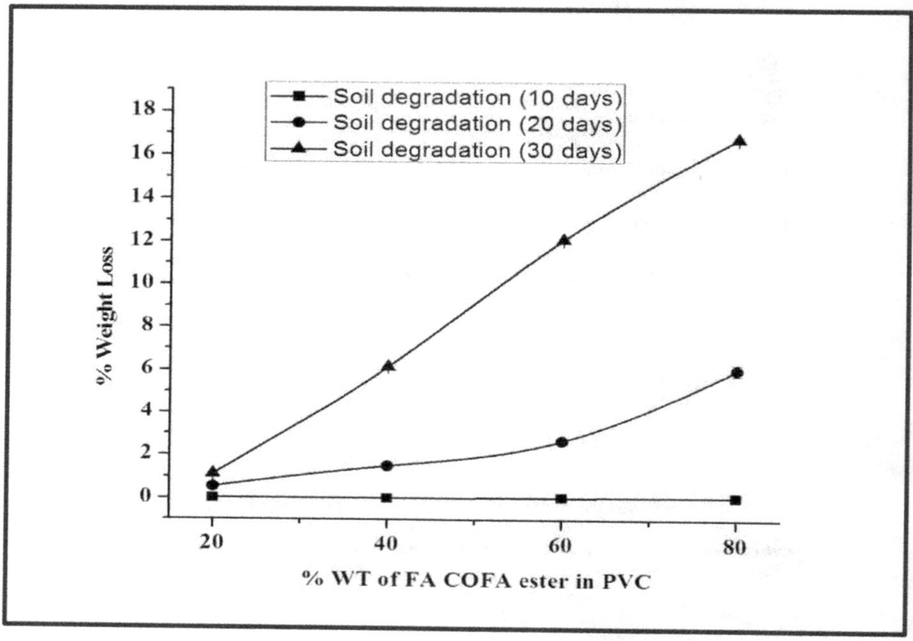

Figure 27.Degradation of PVC+ (DBP:FA-COFA) blends in Soil. Values are represented as mean ± SEM. n = 3.($p<0.05$)

Enzymatic and hydrolytic degradation in phosphate buffer solutions:

The degradation of the PVC + (DBP: FA-COFA) blends films was observed in phosphate buffer and the results are presented in the **Table 42 (vide Figure 28)**. The method is often used for rapid evaluation for the biodegradability test of polymers, though it is an *in-vitro* study and the degradation under natural conditions cannot be revealed through this experiment. With the increasing FA-COFA content within the PVC blend film, the degradation in phosphate buffer increased. At highest FA-COFA concentration(80%w/w), maximum degradation was observed and it was 1.21% after 10 days, 20.51% after 20 days and 47.58 % after 30 days (Figure 28). Hereafter the influence of phosphate buffer was quite evident. The enzymatic degradation of PVC+ (DBP: FA-COFA) films was performed using pancreatic lipase (Figure 28). The enzymatic degradation was found to be lower than the hydrolytic degradation. The PVC film with 80% of FA-COFA ester showed (Figure 28) only 0.13% degradation after 10 days whereas maximum degradation of 3.46% was found to be observed after 30days. Gradual increment in biodegradation was found to be observed with the increasing FA-COFA content in PVC blend and in this case hydrolysis is considered to be the main method of degradation. Generally the hydrolyzed entities are easily consumed by microorganisms to go through further biodegradation. Therefore the hydrolysis was appeared to be the preliminary degradation process followed by microbial degradation in these blend films [123].

Table 42.Enzymatic and hydrolytic degradation behavior of PVC+ (DBP: FA-COFA)blends.

Percentage of FA-COFA ester in PVC(%w/w)	Loss of Weight (%w/w)					
	Hydrolytic (10 Days)	Enzymatic (10 Days)	Hydrolytic (20 Days)	Enzymatic (20 Days)	Hydrolytic (30 Days)	Enzymatic (30days)
20	0.05± 0.33	0.01±0.03	2.04±0.01	0.12 ±0.01	5.94±0.45	1.01±0.05
40	0.15±0.01	0.02±0.04	4.90 ±0.44	0.36±0.01	10.58±0.10	1.67±0.12
60	0.29±0.05	0.05±0.02	11.26±0.15	0.41±0.02	17.99±1.36	2.15±0.06
80	1.22±0.02	0.14±0.02	20.51±0.52	0.6±0.03	**47.58±0.3***	**3.46±0.1#**

Values are represented as mean ± SEM. n = 3 ($p < 0.05$)

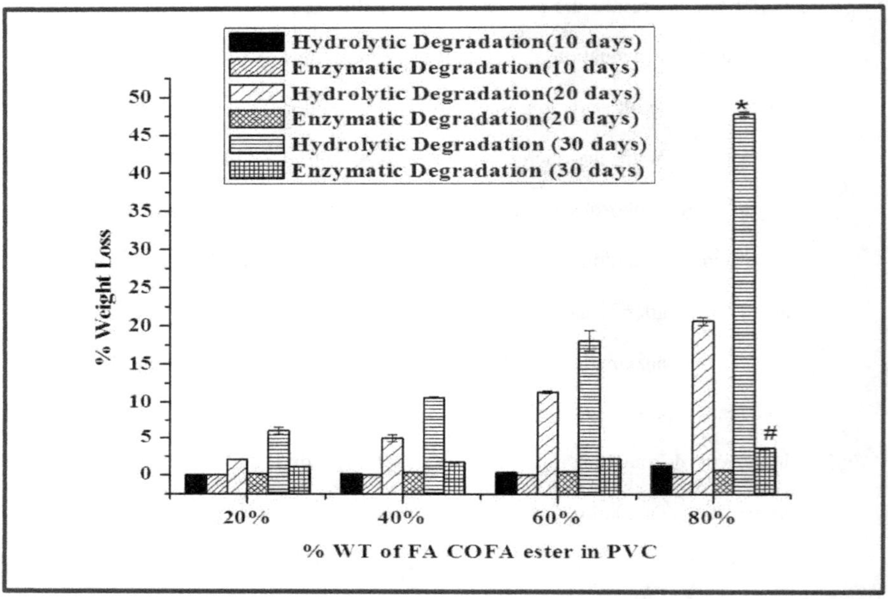

Figure 28.Enzymatic and hydrolytic degradation behavior of PVC+ (DBP :FA-COFA) blends. Values are represented as mean ± SEM. n = 3($p < 0.05$)

***# Superscripts in the different bars indicate significant differences ($p < 0.05$)**

PERFORMANCE EVALUATION OF THE FA-COFA ESTER PLASTICIZED EC FILMS

X Ray diffraction analysis:

Figure 29 depicts XRD patterns of pure EC film, EC film containing different concentration (%w/w) of FA-COFA ester as plasticizer (EC + FA-COFA5, EC + FA-COFA10, EC + FA-COFA15, EC + FA-COFA20,EC + FA-COFA25). Di-butyl phthalate was used as conventional plasticizer at different concentration in the EC film (EC + DBP 5, EC + DBP15 and EC + DBP25). Pure EC film was amorphous whereas in case of EC + FA-COFA 5 the peaks around $10^{0}$and 20^{0}wasenhanced. Therefore the addition of FA-COFA ester as plasticizer enhanced the crystalline property of the EC film. As the concentration of the FA-COFA ester increased (up to EC + FA-COFA 15) in the EC film, sharp diffraction peaks were observed at $10^{0}$and 20^{0}. But in case of EC + FA-COFA20 and EC + FA-COFA25 the diffraction peaks at$10^{0}$and 20^{0}were decreased. When DBP was applied in the EC film as conventional plasticizer, highest diffraction peaks at $10^{0}$and $20^{0}$was observed in case of EC + DBP 5 which were similar to EC + FA-COFA 5. But in case of EC + DBP 15 and DBP 25 the diffraction peaks were decreased, whereas in case EC + FA-COFA 5, EC+ FA-COFA10and EC + FA-COFA 15 the diffraction peaks were increased.

Hence, with increment of FA-COFA ester at a certain concentration (up to EC + FA-COFA 15), the degree of crystallization of amorphous EC film increased.

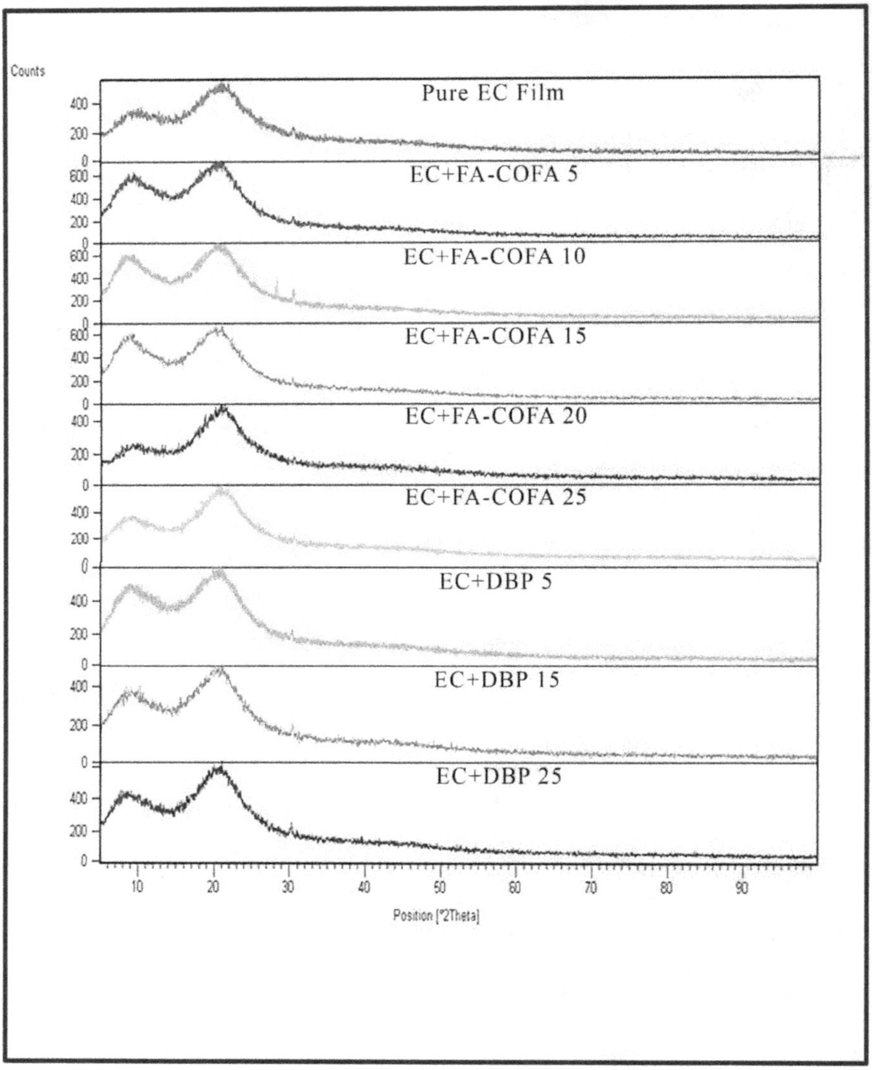

Figure 29. X-ray diffraction analysis of Pure EC film, EC film containing FA-COFA ester at different concentration i.e. 5,10,15,20 and 25%w/w and the EC film containing DBP as conventional plasticizer at different concentration i.e. 5,15, and 25%w/w

Surface morphology:

Figure 30 (a,b,c,d,e and f) and Figure 31 (a,b and c) showed the SEM and AFM images of the EC film surface respectively.

Scanning Electron Microscopy:

SEM (Figure 30 b) image of the EC film containing 5% of FA-COFA ester (EC + FA-COFA5) displayed a smooth surface which was similar to the surface of pure EC film (Figure 30 a). Therefore, the FA-COFA ester was miscible with EC and even distribution was observed in the SEM image (Figure 30 b) of the EC + FA-COFA5. At 15% concentration of the FA-COFA ester (EC + FA-COFA15), the SEM image (Figure 30 c) of the EC film showed smooth surface, whereas in case of EC + DBP15 film the surface was quite rough. Hence, it could be said that FA-COFA ester as plasticizer was more compatible with EC in comparison to DBP (Figure 30 e). SEM images of the EC film containing 25% of the FA-COFA ester (Figure 30 d) and DBP showed (Figure 30 f) rough surface, probably the higher concentration may lead to the aggregation of plasticizer with EC film.

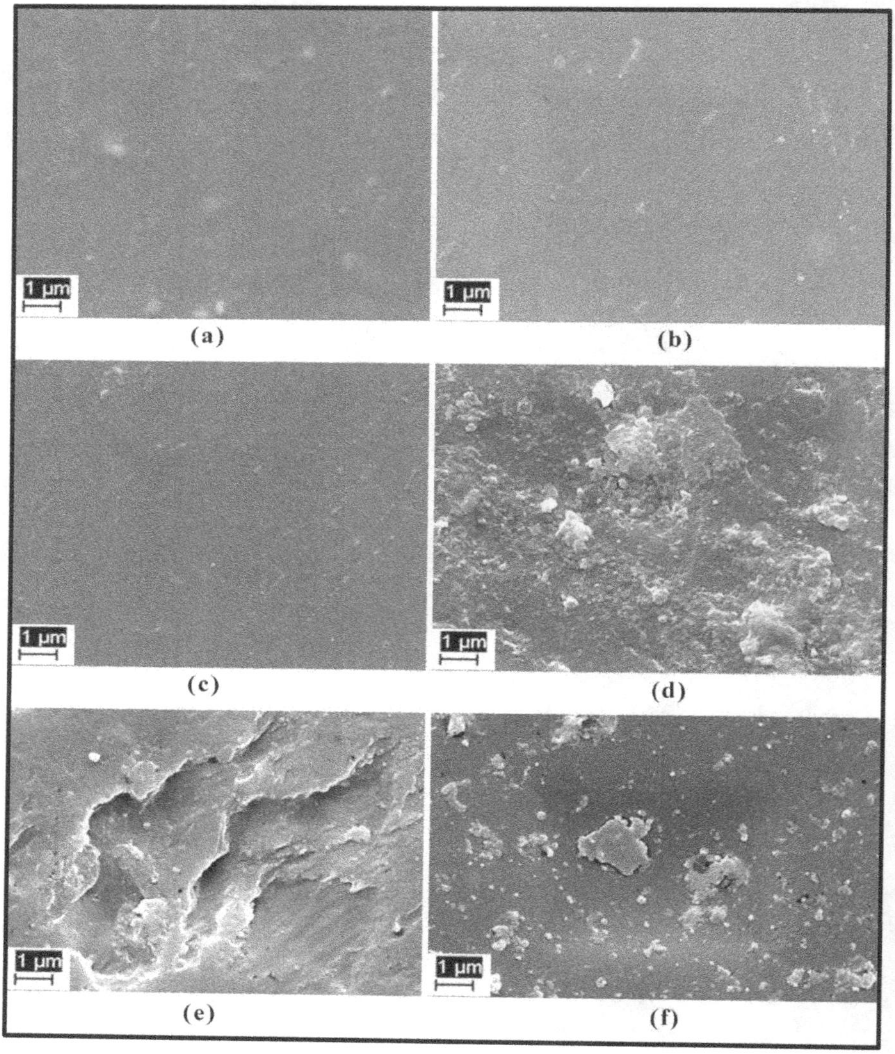

Figure 30. SEM images of EC film surfaces (a) Pure EC film, (b) EC + FA-COFA5 film, (c) EC + FA-COFA15 film , d) EC + FA-COFA25 film, (e) EC + DBP15 film, (f) EC + DBP25 film

Atomic Force Microscopy:

The AFM images of topography and 3D topography were listed in **Figure 31(a,b and c)** for the investigation of surface structure of pure EC film, FA-COFA ester and DBP plasticized EC film respectively. From topography and 3D topography images (Figure 31 a) of pure EC film, it was observed the film surface was embraced of smooth surface with some nodular structure. Such structures were also observed in a previous study of the EC [110]. In case of the film of EC + FA-COFA 15(EC film with 15% w/w FA-COFA ester as plasticizer), the nodules were higher in number (Figure 31 b) and evenly distributed over the film surface, whereas EC + DBP 15(EC film containing 15% DBP as conventional plasticizer) showed uneven distribution of nodules over the film surface (Figure 31 c). This difference indicated the influence of plasticizers on the distribution of molecule over the film surface. The polymer chains of the EC were compelled to close proximity, when the EC solution was cast into the petri dish and the solvent got evaporated. Hence, the plasticized film containing specific functional group or long chains could influence the formation of several bonds, such as hydrogen bond, di-pole di-pole bond etc.

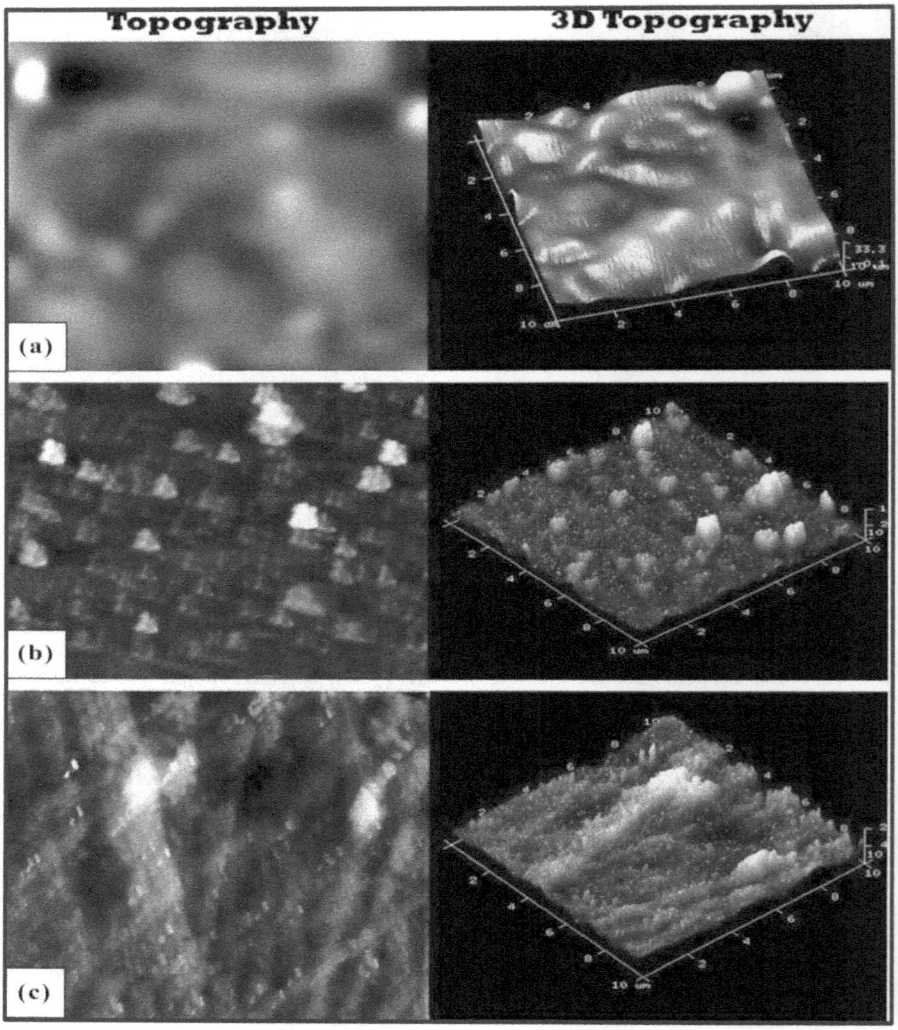

Figure 31.AFM images of EC film surfaces (a) Pure EC film, (b) EC + FA-COFA 15 film, and (c) EC + DBP 15 film

SEM and AFM images showed different surface morphologies of the films due to different molecular weight, hydrophobicity and polarity of the plasticizers.

Mechanical performances:

Table 43 illustrated the mechanical behavior of the EC films such as the tensile strength, Young's modulus strength and the percent-age of elongation at its break. At the higher concentration of the FA-COFA ester, EC film had a lower tensile property and Young' s modulus strength. The tensile strength (Table 43) decreased from33.21 MPa (Pure EC film) to 7.90–11.65 MPa (at 25% of the FA-COFA concentration), whereas in case of DBP which was a conventional plasticizer the tensile property decreased to 14.15 MPa (at 25% concentration) only. At 5% concentration of DBP (EC + DBP 5), the tensile strength was increased in comparison to the pure EC film (Table 43).Therefore, the EC film containing FA-COFA ester showed better the tensile property than the films containing DBP as plasticizer. The Young's modulus strength decreased from 877 MPa (Pure EC film)to 397.03 MPa (EC + FA-COFA 25), whereas in case of EC + DBP 25the Young's modulus strength decreased to 693.61 MPa only [110]. Hence the FA-COFA ester containing EC films also showed better result in case of Young's modulus strength than DBP containing EC films. At the higher concentration of the FA-COFA ester, the ester molecules might have better impact on the interactions among neighboring EC Chains. This resulted in a significant decrease in the tensile strength and Young's modulus strength. DBP containing films also showed (Table 43) decrease in the tensile strength and Young's modulus strength but it did not exceed the performance e of FA-COFA ester containing films, that might be due to the lower molecular weight of DBP. Table 43 also showed the elongation property of the EC films containing both FA-COFA ester and DBP as plasticizer. The elongation property of the EC film containing FA-COFA ester showed almost similar performance when compared with the performance of the EC film containing DBP as plasticizer [110]. With the increasing concentrations of the DBP plasticizer in the EC film, there was a decrease in the elongation property. In case of FA-COFA ester containing EC film, the elongation property decreased to 15.11%for EC + FA-COFA 5 film. But the films containing FA-COFA ester at the concentration of 10% and 15%, there was increase (17.46%for EC + FA-COFA 10 and 20.11% for EC + FA-COFA 15) in the elongation property and this property was similar to the performance of the conventional plasticizer DBP. Thus the FA- COFA ester not only contained good tensile strength and modulus elasticity but also consisted of similar elongation property as the conventional plasticizer such as DBP.

Table 43.Mechanical performance of EC film

Sample	Young's Modulus (MPa)	Tensile Strength(MPa)	Elongation at break (%)
EC	877±0.52	33.21 ±0.73	16.03±0.36
EC+FA-COFA 5	1327.76±1.09	29.12±0.24	15.11±0.35
EC+FA-COFA 10	707.96±1.59	23.86±0.14	17.46±0.13
EC+FA-COFA 15	661.55±0.99	21.37±0.31	**20.19±0.28****
EC+FA-COFA 20	495.61±0.93	**7.90±0.21**$^{\#}$	10.56±0.08
EC+FA-COFA 25	**397.03±2.09***	11.65±0.23	8.45±0.10
EC+DBP 5	1353.53±3.23	38.73±0.32	15.858±0.07
EC+DBP15	876.41±1.19	21.17±0.19	19.472±0.37
EC+DBP 25	693.61±2.03	14.15±0.27	11.41±0.14

Values are represented as mean ± SEM. n = 3.Superscripts in the different column indicate significant differences ($p < 0.05$)

Thermal behavior :

Figure 32 depicted the TGA curves of the pure EC film and FA-COFA plasticized EC films. Pure EC film exhibited a good thermal stability and the film dissociation started at around 300°C.EC films containing FA-COFA ester (at the concentration of 15%)showed almost same thermal stability as the pure EC film but a loss of weight was observed at little less temperature. This phenomenon might occur due to the escape of ester from the film along with the degradation of EC. Similarly, EC film with higher concentration of FA-COFA ester (25%) exhibited (Figure 32) a little more deterioration was observed in the thermal stability as well as weight lossin comparison to the pure EC. This incident might occur due to the faster escape of ester in case of EC + FA-COFA25 when compared with EC + FA-COFA15; though the difference between two concentrations was quite insignificant. At higher temperature, TGA curves of both the ester plasticized EC films (EC + FA-COFA15 and EC + FA-COFA25) showed (Figure 32) elevations in the thermograms while after a certain temperature rise of around 280–300°C, the films began to degrade. Whereas, in case of EC film containing DBP as conventional plasticizer a relatively poor thermal stability was observed in a previous literature [110,147]. Degradation of the EC film started at 160°C, along with the DBP escape.

Hence, the EC film containing 15% FA-COFA ester (EC + FA-COFA15) as plasticizer exhibited almost similar thermal stability as the pure EC film, which proved that FA-COFA ester at a certain concentration can be an efficient and safer option for the plasticizer industry.

Figure 33 illustrated the DSC curves of the EC films, FA-COFA plasticized EC film and DBP plasticized EC films. Pure EC film showed good thermal stability. Endotherm of EC film containing FA-COFA ester (at the concentration of 15%) showed similar stability as the pure EC film. Thus the FA-COFA containing film showed good stability in presence of oxygen at high temperature.

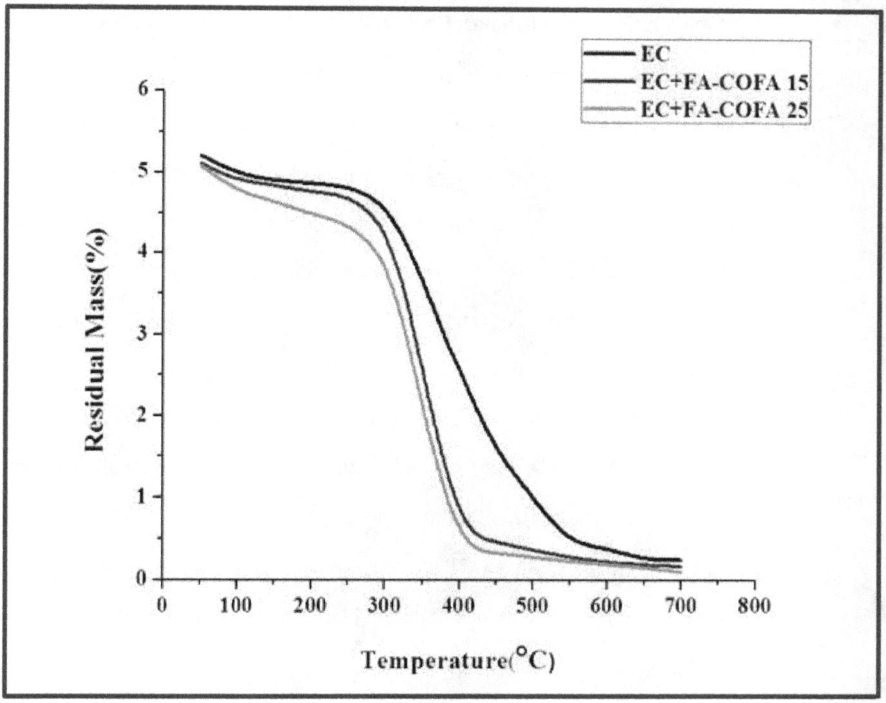

Figure 32.TGA curves of EC films with or without FA-COFA ester as plasticizer

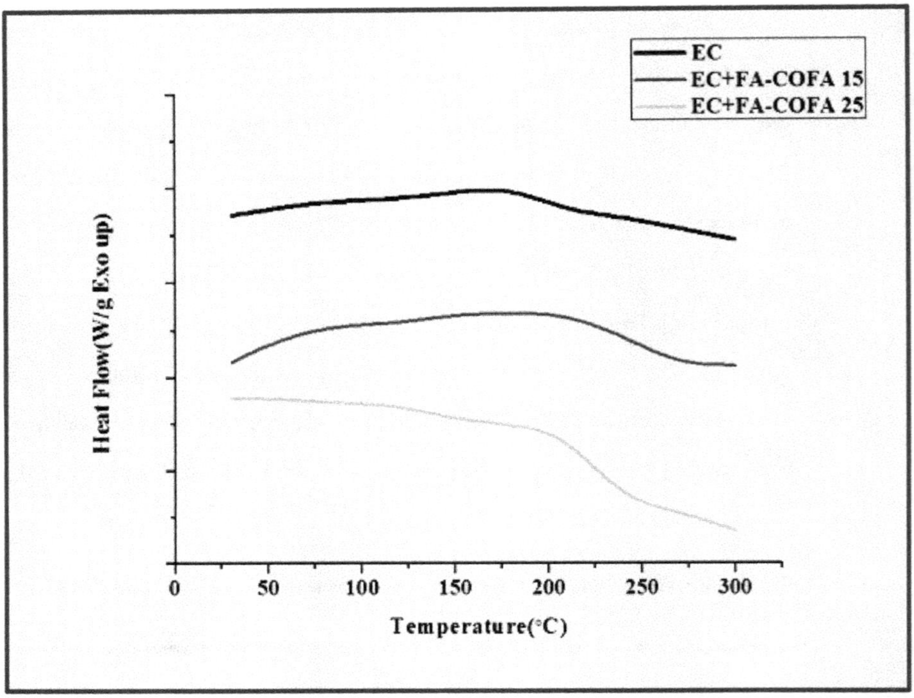

Figure 33.DSC curves of EC films with or without FA-COFA ester as plasticizer

Water vapor permeability properties

Water vapor permeability (WVP) has great importance in food and pharmaceuticals industries mainly used for food coating and packaging films [110,147]. **Table 44** described the water vapor permeability of both pure and plasticized EC films. The WVP of the EC films diminished with increasing concentration of both the FA-COFA ester and DBP. But the better results were found at higher concentration of FA-COFA ester containing EC films(EC + FA-COFA) in which decreased WVP was observed (Table 44). It might be due to the compactness of EC and FA-COFA ester at higher concentration. EC + DBP films also showed (Table 44) lower WVP in comparison to pure EC film. Being a hydrophobic plasticizer, the DBP in higher concentration in the EC film (EC + DBP25) showed (Table 44) higher WVP in comparison to EC + DBP 5. This phenomenon might have occurred due to the diffused structure formed between EC and DBP. Precisely, it can be said that EC film, containing FA-COFA ester as plasticizer showed lower WVP which is an essential property for coating and food packaging industries.

Table 44.Composition and water vapor permeability of EC film with or without plasticizer

Sample	EC(g)	Plasticizer (FA-COFA ester/DBP)	Water Vapor permeability (10^{-5}g/cm^2/24 h)
EC	2	0	1.85±0.02
EC+FA-COFA5	1.9	0.1	1.75±0.03
EC+FA-COFA 10	1.8	0.2	1.59±0.01
EC+FA-COFA 15	1.7	0.3	1.50±0.05 [a]
EC+FA-COFA 20	1.6	0.4	1.42±0.02
EC+FA-COFA25	1.5	0.5	**1.39±0.04**
EC+DBP 5	1.9	0.1	1.74±0.02[b]
EC+DBP 15	1.7	0.3	1.72±0.04
EC+DBP 25	1.5	0.5	1.69±0.05

Values are represented as mean ± SEM.n=3. Superscripts in the same column indicate significant differences ($p < 0.05$)

Migration stability tests:

The results of the migration stability tests are shown in **Table 45**. EC film with the addition of FA-COFA ester as plasticizer did not show water migration, indicating hydrophobic nature of these films. Low migration resistance was observed when each of the EC + FA-COFA15 and EC + DBP 15 films was immersed in gasoline and petroleum ether. The migration of the FA-COFA ester was significantly lower into these two solvents in comparison to conventional plasticizer DBP, indicating that the addition of the FA-COFA ester successful in enhancing the plasticizer incorporation into the EC film. Hence FA-COFA ester containing EC film showed better migration stability in comparison to DBP plasticized EC film.

Table 45.Migration of FA-COFA ester and DBP from the EC film into different solvents

Solvent	**Migration(%w/w)**	
	EC+FA-COFA 15	**EC+DBP 15**
Water	-	-
Petroleum ether	**25.56±0.01***	36.78±0.02
Gasoline	**34.56±0.03#**	43.60±0.04

Values are represented as mean ± SEM.n=3 Superscripts in the same column indicate significant differences ($p < 0.05$)

CONCLUSION

The following conclusion emerged from the present study-

- The present study described the successful production of FA-COFA ester by lipase catalyzed esterification reaction between FA and COFA. The ester yield was very significant at 60^0C temperature with 3:1 molar ratio of FA and COFA and 10% (w/w) immobilized lipase concentration for 5 h.
- The ester was found to be partially biodegradable to the environmental exposure up to a certain concentration.
- The concerned FA-COFA ester obtained was successfully incorporated into PVC and ethyl cellulose films prepared by different casting method, showing no migration into water.
- Furthermore the results of the tests in soil burial, enzymatic degradation as well as degradation in buffer showed the polymer blends to be biodegradable. The biodegradability was found to be increased with the increased concentration of FA-COFA content.

SECTION - B

REVIEW OF LITERATURE

Fatty acid esters are important classes of chemical components which have many applications in widespread areas such as plasticizers, pharmaceuticals, solvents, and intermediate for many industries. Esterification reaction of carboxylic acids signifies an extensively used liquid phase reaction for the production of plasticizers, solvents, flavors, and, most importantly, biolubricant and biofuel components. Recently ,a range of fatty acid esters is being produced commercially using immobilized lipase in non-aqueous solvents [139, 155-157] and the concern in industrial practices employing lipase biocatalysts to synthesize more such esters is still growing due to their important and multiple applications. Esters of long-chain fatty acids (12–20 carbon atoms) and short-chain alcohols (3-8 carbon atoms) have been used progressively in the food, detergent, cosmetic and pharmaceutical industries. There are reports on esterification reaction of long-chain acids with long-chain alcohols (12–20 carbon atoms) which have important applications as plasticizers and lubricants [137,139-140].Sperm whale oil, carnauba wax and jojoba oil, have been proposed for such kind of applications but theses oils are expensive and these are not available in bulk amounts[136]. From the economical point of view, it is very much important to develop methods with cheaper raw materials. Vegetable oils are environmentally friendly and are one of the most important classes of renewable resources. Renewable resource-based plasticizers can be used as a platform to replace petroleum based polymers through innovative design[150-153]. New bio-based plasticizers can contend with non-renewable petroleum-based materials on the basis of cost performance with the added advantage of ecofriendliness [120-125].Direct esterification starting from an acid and an alcohol is of great economic importance in the production of renewable materials. Plasticizers based on renewable raw materials and their derivatives are drawing increased attention in various

applications [133]. There are many more reports regarding the synthesis and application of renewable raw material based plasticizers.

Modern enzymology has gained huge improvements in the development and application of lipases as catalysts. New lipase immobilization techniques make it possible to apply enzymes in industrial processes in a same way to the use of classical catalysts for heterogeneous chemical reactions [135-137].Major advantages of using immobilized lipases are, higher yields, working under mild reaction conditions, easy recovery and reusability of the enzyme , besides the different substrate specificity [138-140].The kinetic studies of immobilized lipase catalyzed reactions are comparatively unusual but it has become important in the past few years[158-159,161]. Usually these models are based on the concept of Michaelis-Menten assumptions and seem to be valid for most simple enzymatic reactions [163-165]. According to a report, dehydrogenation of various alcoholic groups using an alcohol-dehydrogenase as catalyst found to be following a kinetic model based on the hypothesis of Michaelis-Menten when the reaction was performed under optimum reaction conditions [133,136-138]. Similar results were found to be observed for the kinetic study of the hydrolysis of starch using a *Rhizopus niveus lipase* immobilized on a ceramic-type support [142]. Researchers suggested a model built on the ping-pong mechanism for the kinetic study of the interesterification of *n*-propyl acetate with geraniol, using lipases as catalysts and different solvents in the reaction media [179]. This hypothesis was accepted in the study of the esterification of lauric acid with menthol using a *Penicillium simplicissimum* [180-181]. Most of the scientists agree that the reaction is conducted via an enzyme-acyl complex when lipases are used as catalysts. This hypothesis was established by studies that showed the possible site for the reaction which is the formation of an active complex between an acyl group belonging to the substrate and one of the active centers of the lipase

[162-170]. It is very much important to consider other physical parameters along with the basic reaction steps. In case of the application of traditional solid catalysts, there is a probability of significant mass transfer due to the dilution of substrates and products, which must be considered in the kinetic model. The most extended methods in this case is based on Langmuir hypothesis[182].Even in some of the studies, it was found to be observed that lipase catalyzed esterification reactions are following the ping-pong bi bi mechanism with enzyme substrate inhibition[158-161].A study on the lipase catalyzed reaction in super critical CO_2 also following ping-pong bi bi mechanism[168].Similar observation was found in case of esterification of furfuryl alcohol with oleic acid [133].Another study regarding the esterification of oleic acid with ethanol showed same kind of observation and following a basic reaction kinetics[172].Rizzi et.al; reported a kinetic study on the lipase catalyzed transesterification of isoamyl acetate in n-hexane[165].Yadav also developed the kinetic model of lipase catalyzed trans esterification of n-octanol with vinyl acetate[173].There is another kinetic study on the trans-esterification of high oleic sunflower oil[174].Kinetic study also reported esterification of free fatty acids derived from waste cooking oil with octanol [125].

Castor oil is environment friendly material for a huge range of industrial applications. As it is inedible, it is an article of commerce. In the search of making renewable resource based plasticizer for medical and commodity plastics to replace phthalate based plasticizer for flexible polymeric materials [128-130]. It is a renewable raw material for coatings, inks, paints, lubricants and an extensive variety of other products. As it is consisting of pendant hydroxyl group, the oil is suitable for synthesis of polyurethane millable [142], castables [175], polyurethane elastomers [176], adhesives and coatings [177], and polyurethane foam [178]. Epoxidized castor oils and castor oil phosphate esters are found to be a good

plasticizers with flame retarded group [129]. Polyol based esters of castor oil are potent environment friendly plasticizers [162].Castor oil based containing THEIC and diethyl phosphate groups are efficient flame retardant plasticizers [128].

Furfuryl alcohol (FA) is a pentosan-rich biomass derived chemical obtained by the hydrogenation of furfural. FA is a biomolecule, which polymerizes readily under acidic conditions and forms a thermo-stable polymer that is widely applied in the foundry industry to produce biorenewable plasticizer. According to a report, Oleic acid was successfully esterified with furfuryl alcohol [133].

MATERIALS AND METHODS

LIPASE CATALYZED ESTERIFICATION REACTION OF FURFURYL ALCOHOL AND CASTOR OIL FATTY ACID

(i)Materials

Pure COFA was procured from Aloka Oil Industry, Kolkata and analyzed at laboratory for fatty acid composition by gas chromatography (GC). FA was obtained from Merck India Ltd., Mumbai, India. *Candida antarctica* lipase B (NS 435 or Novozyme 435), immobilized using macroporous acrylic resin with a water content of 2% (w/w)and particle size of 450 μm was a product of Novozyme India Pvt Ltd. and supplied from Brenntag Ingredients (India) Pvt. Limited, Mumbai, India.

(ii)Method

Esterification was conducted in a flat bottom flask with continuous stirring at 200 rpm under different experimental conditions. Process parameters, studied in present investigation were temperature (T) (45^0C to 65^0C, with $\Delta T = 5^0$C), molar ratio of COFA and FA (M) (1:1 to 1:5 withΔM = 1:1)and immobilized lipase loading (W) (4% to 12% w/w with ΔW = 2% w/w). In order to generate an exhaustive kinetic data set, total 125 experiments were conducted under varied parametric conditions. All experimental runs were continued for 5 h, which was evaluated to be higher than highest time necessary to attain reaction equilibrium for all experimental conditions used in present study.

Reaction coordinate was monitored by determining transient acid values of reaction mixture detail experimental procedure is discussed later. Ester formation was confirmed by thin layer chromatography (TLC). Percent yield of ester and its fatty acid composition was determined simultaneously by GC. Detailed method has been provided in the **Section A.**

Final product was obtained by vacuum distillation under low pressure (28 mm Hg) at 90^0C to remove excess FA. Subsequently, distillate was washed with dilute NaOH solution followed by water to remove free fatty acids. Finally, product was dried under vacuum (28 mm Hg) and stored for further analysis.

PHYSICOCHEMICAL CHARACTERIZATION OF THE ESTER PRODUCED BY THE ESTERIFICATION OF FURFURYL ALCOHOL (FA)AND CASTOR OIL FATTY ACID(COFA)

Determination of acid value:

Detailed method of acid value has been provided in the previous section A.

Fractional conversion of the limiting reactant i.e. COFA can be calculated from the aforementioned acid value as: $X_A = \frac{acid\ value_{t=0} - acid\ value_{t=t}}{acid\ value_{t=0}}$. Additionally, [COFA] was calculated as $[COFA]_t = [COFA]_{t=0}(1 - X_A)$.

Qualitative analysis of esters by Thin Layer Chromatography (TLC):

Detailed method of TLC has been provided in section A.

Analysis of fatty acid composition of COFA and its esters by GC:

Detailed method of GC has been provided in the section A.

KINETIC STUDY

Kinetics of enzymatic esterification can be described by well-known Michaelis-Menten model. However, present reaction system consists of two reactants (FA and COFA) that leads to the formation of ester and water. Accordingly, different substrate binding and product release steps are expected to be associated in the overall reaction scheme of present system. Different modifications of baseline Michaelis-Menten model were proposed and

tested over decades for multi-reactant systems. Among these models, ternary-complex and ping-pong mechanisms are generally recommended [162-170]. Several investigations have confirmed ping-pong mechanism to correctly describe lipase catalyzed esterification [163-165]. Additionally, fatty acid and alcohols are reported to inhibit lipase catalyzed reactions [136-138]. According to baseline ping-pong bi bi scheme, COFA is first expected to bind with lipase, forming an enzyme-substrate complex, which isomerizes to respective acyl-enzyme intermediate with simultaneous release of water. Subsequently, FA reacts with acyl-enzyme complex to form another binary complex, which further isomerizes to lipase-ester complex. Finally, lipase-ester complex breaks into free enzyme and ester. Regarding inhibition pathway, it is imperative to note that both FA and COFA competitively inhibit lipase and acyl-enzyme complex to form dead-end products. Therefore, based on reported observations, present reaction kinetics is best described as:[***Equation 2***]-

$$V = \frac{V_m[COFA][FA]}{K_{m(COFA)}[FA]\left(1+\frac{[FA]}{K_{i(FA)}}\right)+K_{m(FA)}[COFA]\left(1+\frac{[COFA]}{K_{i(COFA)}}\right)+[COFA][FA]} \qquad [2]$$

where V is specific reaction rate (velocity of the reaction), V_m be the maximum reaction rate, $K_{m(.)}$ and $K_{i(.)}$ represent Michaelis constants and inhibition coefficients respectively for two reactants, FA and COFA.

The kinetic parameters of the present system, which include V_m, $K_{m(FA)}$, $K_{m(COFA)}$, $K_{i(FA)}$ *and* $K_{i(COFA)}$ were estimated for each experimental run using non-linear least square regression analysis in MATLAB platform. Moreover, variations of aforementioned kinetic parameters (5 nos) with pertinent experimental parameters, (enzyme loading, W; temperature, T; and molar ratio of the reactants, M) were investigated to decipher the corresponding functional dependence. Knowledge about this kinetic parameter profiles are expected to be highly beneficial in locally maximizing the product yield in cases where the

operation is to carried with allowed navigation over a constraint parametric space, which may be far from global optima.

STATISTICAL ANALYSIS

All the analysis was done in triplicates ($n = 3$). All the data are represented as means with their standard errors (Mean ± SEM). Statistical comparisons among groups were performed using one way ANOVA. Level of significance $p < 0.05$ was considered by Tukey test(using Origin Pro 8 software).The %conversion was analyzed by the RSM using the software Design Expert, Version 7.0 Minneapolis, USA. The kinetic parameters of the present system were estimated for each experimental run using non-linear least square regression analysis in MATLAB platform.

RESULTS AND DISCUSSION

KINETIC PROFILE AND PARAMETRIC DEPENDENCE

To understand the gross qualitative trend of the governing kinetics of biocatalytic esterification reaction between FA and COFA ,it is necessary to explore the transient concentration profiles of COFA under varied parametric conditions of (i) reactant's molar ratio (dimensionless) (ii) enzyme loading (% w/w) (iii) temperature (^{0}C).The corresponding trends are depicted in **Table 46,47 and 48(vide Figure 34 a, b and c)** respectively. **Table 46,47 and 48(vide Figure 34 a, b and c)** represent progressively decreasing [COFA] before leveling off to some characteristic equilibrium value as COFA is being consumed by the FA. Moreover, the rate of decrease i.e$-\frac{d[COFA]}{dt}$, which is equivalent to the specific reaction rate, was recorded to decrease with time, as evident. **Table 46(vide** Figure 34a) represents the effect of molar ratio on the transient COFA concentration over the time interval of 0 to 5 h under otherwise fixed parametric conditions (temperature: 60 °C and enzyme loading:10% w/w). Over the aforementioned batch time, the highest decrease in COFA concentration was recorded at a molar ratio of 3:1. **Table 47 (vide** Figure 34b) described the effect of enzyme loading under constant reaction temperature and molar ratio (60°C and 3:1 respectively). The figure showed highest decrease of concentration of COFA was also observed at 10% w/w enzyme loading over 5 h of reaction time. It is worth noting that with the increasing enzyme loading, the gross depletion of COFA concentration was monotonically increasing, however, as the enzyme loading was increased from 10% to 12% (w/w), the depletion was observed to decrease .This may be attributed to the probable enzyme-substrate inhibition. According to a report, this phenomenon might have happened due to the steric hindrance which is produced by excessive enzyme concentration [136-137].**Table 48 (vide** Figure 34c)

represents the impact of reaction temperature (45^0-65^{0}C) on the transient acid value profile under constant conditions of molar ratio of FA and COFA (3:1) and enzyme loading (10% w/w). Maximum depletion of acid value was observed at 60^{0}C.With the increase of temperature from 60^{0}C to 65^{0}C, the gross depletion of acid value was noted to decrease slightly, which may be contributed due to the enzyme inactivation at temperature, above certain system-specific optimum value.

Table 46.Changes in concentration of COFA at different molar ratios (1:1 to 5:1) of FA and COFA using NS 435 as biocatalyst (10%w/w) at different time intervals (0 h to 5 h) and at 60^{0}C reaction temperature

Time(h) ↓	Concentration of COFA(mol L^{-1})				
	Molar Ratio(FA:COFA)				
	1:1	2:1	3:1	4:1	5:1
0	2.48±0.1	2±0.13	1.7±0.2	1.5±0.1	1.2±0.1
1	1.41±0.2	1.03±0.2	0.31±0.15	0.40±0.15	0.47±0.2
2	0.76±0.15	0.61±0.2	0.12±0.2	0.21±0.2	0.35±0.15
3	0.66±0.01	0.55±0.25	0.07±0.3	0.15±0.23	0.24±0.25
4	0.45±0.2	0.35±0.15	0.03±0.35	0.11±0.15	0.20±0.2
5	0.34±0.25	0.18±0.1	**0.02±0.6***	0.08±0.2	0.17±0.15

Values are mean ± SEM. $p<0.05$, n=3

Table 47.Changes in concentration of COFA at different enzyme loading(%w/w)of NS 435 at a fixed molar ratio of FA and COFA (3:1) at different time intervals (0 h to 5 h) and at 60°C reaction temperature

Time(h) ↓	Concentration of COFA(mol L^{-1}) Enzyme loading(%w/w) 4	6	8	10	12
0	1.7±0.03	1.7±0.20	1.7±0.10	1.7±0.20	1.7±0.20
1	0.81±0.02	0.71±0.15	0.66±0.03	0.34±0.30	0.41±0.15
2	0.6±0.10	0.46±0.25	0.41±0.2	0.16±0.10	0.20±0.20
3	0.30±0.15	0.25±0.10	0.21±0.025	0.07±0.15	0.16±0.10
4	0.24±0.2	0.19±0.30	0.15±0.30	0.03±0.30	0.12±0.20
5	0.20±0.3	0.15±0.20	0.11±0.10	**0.02±0.20***	0.06±0.30

*Values are mean ± SEM. p<0.05, n=3

Table 48. Changes in concentration of COFA at different reaction temperature (45-65°C)using NS 435 as biocatalyst (10%w/w) at a fixed molar ratio of FA and COFA (3:1) at different time intervals (0 -5 h)

Time(h) ↓	Concentration of COFA(mol L^{-1}) Temperature(° C) 45	50	55	60	65
0	1.7±0.12	1.7±0.30	1.7±0.12	1.7±0.30	1.7±0.12
1	0.72±0.11	0.62±0.21	0.53±0.21	0.34±0.21	0.40±0.21
2	0.42±0.05	0.33±0.15	0.26±0.15	0.12±0.15	0.17±0.15
3	0.34±0.15	0.25±0.11	0.18±0.11	0.06±0.11	0.11±0.11
4	0.30±0.12	0.20±0.20	0.14±0.18	0.02±0.20	0.09±0.18
5	0.20±0.11	0.13±0.10	0.07±0.20	**0.01±0.10***	0.05±0.20

Values are mean ± SEM. p<0.05, n=3

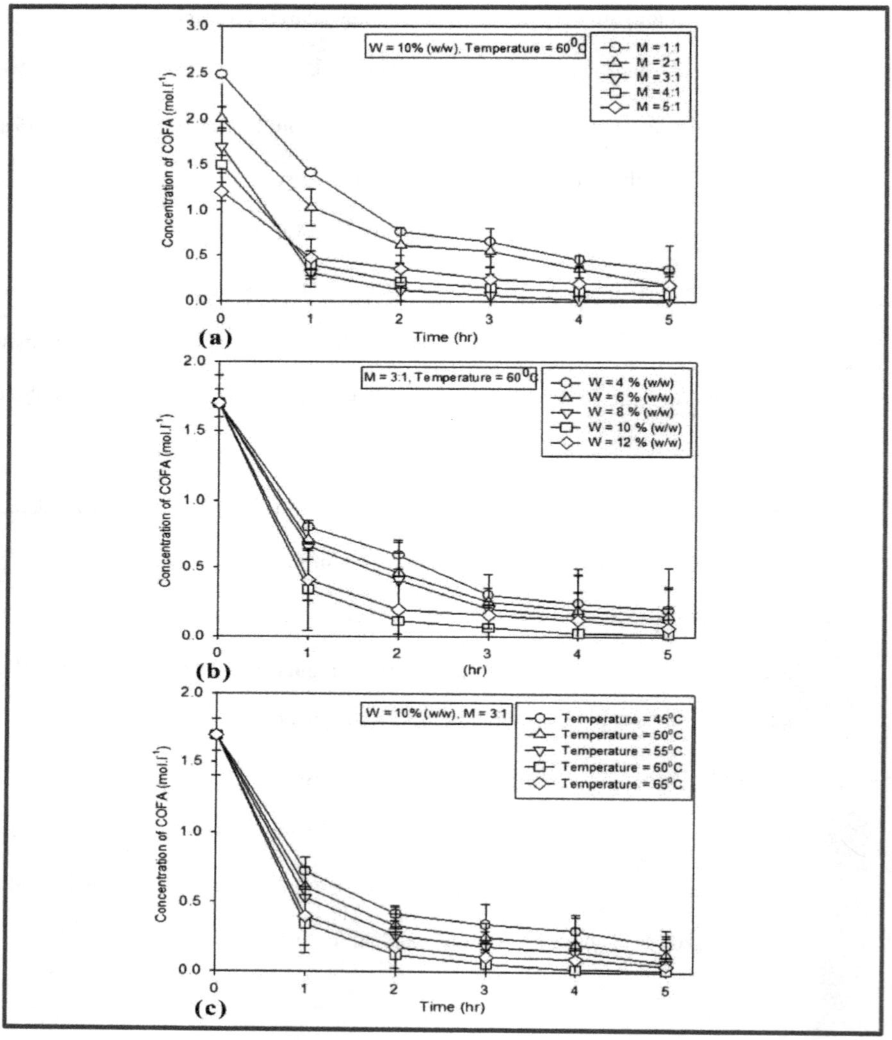

Figure 34.Kinetic profile and parametric dependence of the esterification reaction of FA and COFA: (a, b and c) Changes in concentration of COFA in the course of reaction with variable reaction parameters such as (a) Molar ratio,(b) enzyme loading(%w/w) and (c) temperature (^{o}C) .Values are mean ± SEM. $p<0.05$, $n=3$

To be more informative regarding the exiting kinetic trend, the initial rate, V_0 was calculated under different parametric conditions, which are shown in **Table 49,50,51,52,53 and 54 (vide Figure 35 a-I, a-II, b-I, b-II and c-I, c-II). Table 49 and 50 (vide Figure 35 a-I and a-II)**, represent the variation of the initial rate (V_0) with molar ratio at different reaction temperature (45°C to 65°C) and enzyme loading (4 to 12% (w/w) respectively. **Table 49 (vide** Figure 35a-I), represents the effect of different temperature on the initial rate (V_0) of the reaction over the molar ratio of FA and COFA (1:1 to 3:1),under the fixed parametric conditions (Enzyme loading:10% w/w and time : 5h).Higher value of V_0 was observed in case of 3:1 molar ratio of FA and COFA at all the temperatures(45 to 65°C) but the highest value was observed in case of 60°C temperature. **Table 50(vide** Figure 35a-II) depicted the effect of enzyme loading on the initial rate (V_0) of the reaction over the molar ratio of FA and COFA under the constant reaction parameters (Temperature 60°C and time 5h). Some exception found in both **Table 49 and 50 (vide Figure 35 a-I and a-II)** where 4:1 and 5:1 molar ratio of FA and COFA showed decreased value for V_0 in all the reaction temperatures (45°-65°C) and enzyme loads (4%-12% w/w) respectively. This phenomenon might be caused due to the enzyme –substrate inhibition of the reaction which may be caused by the steric hindrance. From, **Table 49 and 50(vide Figure 35 a-I and a-II)**,it can be said that 3:1 molar ratio of FA and COFA was the optimum for this reaction.

Table 49.Initial reaction rates (V_0) as a function of different reaction temperatures (45-65°C) at different molar ratio of FA and COFA (1:1 to 5:1) using NS435 as biocatalyst (10% w/w)

Molar Ratio(FA:COFA)	V_0(mol$L^{-1}h^{-1}$) Temperature(° C)				
	45	50	55	60	65
1:1	0.72±0.12	0.82±0.11	0.89±0.12	1.02±0.3	0.95±0.12
2:1	0.82±0.11	0.89±0.12	0.98±0.21	1.06±0.21	1.02±0.11
3:1	1.03±0.05	1.11±0.05	1.19±0.11	**1.30±0.15***	1.24±0.05
4:1	0.97±0.15	1.02±0.12	1.10±0.12	1.20±0.05	1.15±0.14
5:1	0.77±0.12	0.81±0.12	0.83±0.18	0.89±0.12	0.86±0.11

Values are mean ± SEM. $p<0.05$, n=3

Table 50. Initial reaction rates (V_0) as a function of different Enzyme loading (4-12%w/w) at different molar ratio of FA and COFA(1:1 to 5:1) at 60°C reaction temperature

Molar Ratio(FA:COFA)	V_0(mol$L^{-1}h^{-1}$) Enzyme loading(%w/w)				
	4	6	8	10	12
1:1	0.74±0.05	0.81±0.11	0.87±0.14	1.01±0.12	0.94±0.11
2:1	0.84±0.11	0.92±0.21	0.98±0.11	1.15±0.03	1.08±0.14
3:1	0.99±0.12	1.08±0.10	1.15±0.03	**1.30±0.14***	1.24±0.16
4:1	0.93±0.15	1.01±0.14	1.07±0.18	1.18±0.21	1.14±0.15
5:1	0.86±0.12	0.93±0.11	0.99±0.16	1.08±0.15	1.03±0.05

Values are mean ± SEM. $p<0.05$, n=3

Table 51 and 52 (vide Figure 35b-I and b-II)depicted the impact of different enzyme load (4% to 12% w/w) and different molar ratios (1:1 to 5:1) of FA and COFA respectively on the reaction temperature ranging from 45°-65°C at a constant reaction time (5h).From the Figure 35(b-I and b-II),highest V_0 was observed at 10% w/w enzyme load and 3:1 molar

ratio of FA and COFA at 60°C of reaction temperature. **Table 51 (vide** Figure 35b-I) showed highest value of V_0 at 60°C reaction temperature in comparison to other reaction temperatures(45^0,50^0,55^0 and 65^{0}C) in all the cases of enzyme loading (4%-12% w/w) at 3:1 molar ratio of FA and COFA and at 12%(w/w) enzyme loading showed an initial rates. But in each case of enzyme loading **Table 51 (vide** Figure 35 b-I),there was a trend of fall in the initial rate at 65°C temperature which might be happened due to the enzyme inactivation at higher temperature. Furthermore the decrease in V_0 was also observed at the same temperature. **Table 52 (vide** Figure 35b-II) initial reaction rates as a function of different molar ratio of FA and COFA(1:1 to 5:1) at different reaction temperature (45-65^{0}C) at an enzyme loading of 10%w/w. **Table 52 (vide** Figure 35b-II)showed highest value of V_0 at 3:1 molar ratio of FA and COFA in comparison to other molar ratios (1:1,2:1,4:1 and 5:1) in all the cases of reaction temperatures (45-65^{0}C) at 10%w/w enzyme loading. But a decrease in the initial rate was observed at 4:1 and 5:1 molar ratio of FA and COFA in each case of reaction temperature (45-65^{0}C).This might occurred due to the enzyme-substrate inhibition caused by the excess amount of FA being added to the reaction system. There are lots of studies found to be observed in support of these phenomena [130-135].

Table 51. Initial reaction rates (V_0) as a function of different Enzyme loading (4-12%w/w) at different reaction temperature(45-65^{0}C) at a molar ratio of FA and COFA (3:1)

Temperature(^{o}C)	V_0(molL^{-1}h^{-1})				
	Enzyme loading(%w/w)				
	4	6	8	10	12
45	0.79±0.10	0.85±0.10	0.9±0.09	1.03±0.10	0.97±0.10
50	0.83±0.06	0.89±0.03	0.95±0.08	1.11±0.05	1.03±0.08
55	0.86±0.07	0.94±0.06	1.03±0.15	1.19±0.06	1.12±0.06
60	0.92±0.03	1.01±0.08	1.12±0.07	**1.30±0.10***	1.23±0.08
65	0.86±0.10	0.97±0.09	1.09±0.056	1.24±0.05	1.20±0.05

Values are mean ± SEM. $p<0.05$, n=3

Table 52. Initial reaction rates (V_0) as a function of different molar ratio of FA and COFA(1:1 to 5:1) at different reaction temperature (45-65^{0}C) at an enzyme loading of 10%w/w

Temperature(^{o} C)	V_0(mol$L^{-1}h^{-1}$)				
	Molar Ratio(FA:COFA)				
	1:1	2:1	3:1	4:1	5:1
45	0.82±0.10	0.88±0.03	1.01±0.06	0.92±0.08	0.97±0.11
50	0.86±0.06	0.92±0.08	1.08±0.03	0.97±0.03	1.02±0.07
55	0.94±0.07	1.01±0.05	1.18±0.07	1.06±0.05	1.12±0.05
60	1.01±0.03	1.1±0.11	**1.30±0.05***	1.18±0.11	1.25±0.08
65	0.95±0.06	1.02±0.03	1.23±0.1	1.09±0.11	1.16±0.03

Values are mean ± SEM. $p<0.05$, n=3

Table 53 and 54 (vide Figure 35c-I and c-II)described the effect of different temperature (45° to 65°C) and different molar ratios (1:1 to 5:1) of FA and COFA respectively on the enzyme loading ranging from 4% to 12% w/w at a constant reaction time (5h).From the**Table 53 and 54 (vide Figure 35c-I and c-II)**,highest V_0 was observed at 60°C reaction temperature, at 3:1 molar ratio of FA and COFA at 10% (w/w) enzyme loading. **Table 53 (vide** Figure 35c-I) showed highest value of V_0 at 10% w/w enzyme loading at 60^{0}C temperature in comparison to other enzyme loading (4%,6%,8% and 12% w/w) at a constant molar ratio (3:1).But the fall was observed in the initial rate of the reaction at 65^{0}C and it was observed in all the cases of enzyme loading (4-12%w/w).This might occurred due to the enzyme inactivation due to the higher temperature. Figure 35(c-II) showed highest value of V_0 at 3:1 molar ratio of FA and COFA at a constant reaction temperature (60^{0}C).But a decrease in the initial rate was observed at 4:1 and 5:1 molar ratio of FA and COFA in each

case of Enzyme loading (4-12%w/w) at 60°C.This might occurred due to the enzyme-substrate inhibition caused by the excess amount of FA in the reaction system .

Table 53. Initial reaction rates (*V_0*) as a function of different reaction temperature(45-65°C) with different Enzyme loading (4-12%w/w) at a molar ratio of FA and COFA (3:1)

Enzyme loading(%w/w)	V_0(molL^{-1}h^{-1}) Temperature(°C)				
	45	50	55	60	65
4	0.80±0.11	0.83±0.10	0.86±0.08	0.94±0.12	0.9±0.05
6	0.82±0.14	0.86±0.05	0.91±0.11	0.98±0.06	0.94±0.12
8	0.91±0.10	0.97±0.12	1.02±0.15	1.12±0.07	0.94±0.15
10	1.05±0.07	1.11±0.15	1.19±0.2	**1.30±0.14***	1.24±0.11
12	0.94±0.05	1.01±0.13	1.07±0.05	1.27±0.10	1.17±0.06

Values are mean ± SEM. p<0.05, n=3

Table 54. Initial reaction rates (*V_0*) as a function of different molar ratio of FA and COFA (1:1 to 5:1) at with different Enzyme loading (4-12%w/w) at 60°C reaction temperature

Enzyme loading(%w/w)	V_0(molL^{-1}h^{-1}) Molar Ratio(FA:COFA)				
	1:1	2:1	3:1	4:1	5:1
4	0.75±0.10	0.8±0.10	0.93±0.09	0.89±0.10	0.85±0.10
6	0.84±0.05	0.87±0.03	0.98±0.08	0.95±0.05	0.92±0.08
8	0.96±0.08	1.01±0.06	1.15±0.15	1.11±0.06	1.06±0.06
10	1.08±0.08	1.14±0.08	**1.30±0.07***	1.22±0.10	1.18±0.08
12	0.98±0.09	1.02±0.09	1.20±0.06	1.12±0.05	1.07±0.05

Values are mean ± SEM. p<0.05, n=3

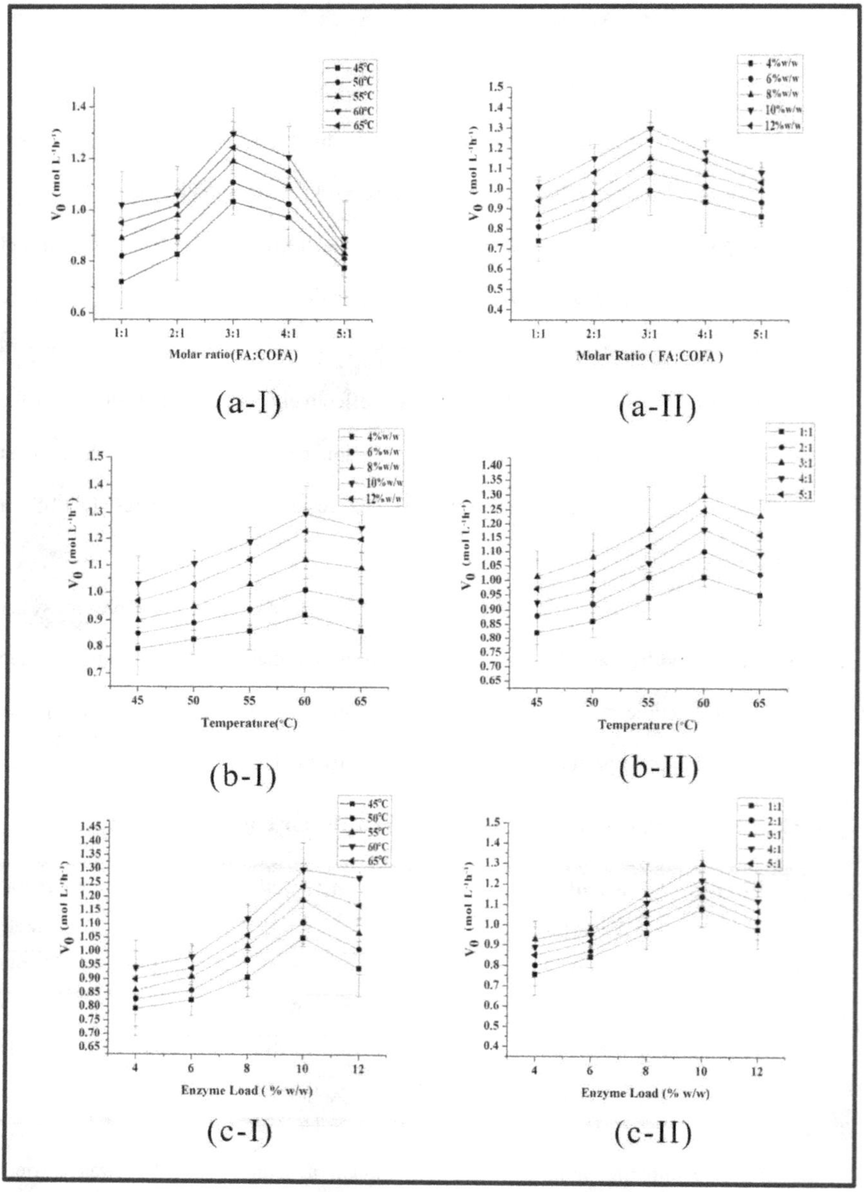

Figure 35. (a-I,a-II,b-I,b-II,c-I and c-II) Initial reaction rates as a function of different molar ratio of FA and COFA (1:1 to 5:1), different Enzyme loading (4-12%w/w) and different reaction temperature(45-65^{0}C). Values are mean ± SEM. $p<0.05$, $n=3$

Optimizing the process parameters

To maximize the % conversion of COFA with respect to the three most pertinent process parameters, namely, (i) molar ratio (FA: COFA, dimensionless) (ii) temperature (in ^{0}C) and (iii) enzyme load (w/w %) standard Response Surface Methodology (RSM) was employed in the present investigation. Accordingly, the standard regression models, which include (i) linear, (ii) 2FI (iii) quadratic, and (iv) cubic regression, were tested to represent the response (% conversion) as a function of aforementioned process parameters (regressor variables). Primarily, the optimum model, which most effectively represent the functional form was selected upon comparing the standard regression parameters, which include (i) standard deviation, (ii) R^2 and adjusted R^2, (iii) predicted residual sum of squares (PRESS), and (iv) the$p$-value for lack of fit.

The results are summarized in **Table 55.** The table clearly indicates the present system to be best described by the cubic model. It is worth noting that instead of inherent mathematical complexity, cubic model was selected because of its much superior regression performance in representing the present data set relative to other models.

Table 55.Comparison between standard models (for esterification)

Source	Standard Deviation	R^2	*Adjusted* R^2	PRESS	p-Value for lack of fit
Linear	5.02	0.5291	0.5167	3133.2	<0.0001
2FI	4.40	0.6484	0.6294	2461.11	<0.0001
Quadratic	4.02	0.7139	0.6901	2038.67	<0.0001
Cubic	1.72	0.9524	0.9432	436.31	<0.0001

After selecting the optimum regression model, a detailed ANOVA was carried out to evaluate the effects of each parameter. The results are summarized in the supplementary **Table 56.** High F-value (= 103.28) of the model confirms that there is only a 0.01% chance

that an F-value this large could occur because of noise [135]. Regarding the significance of model terms, it may be noted that all the linear terms, except T (temperature) i.e. M (molar ratio) and W (catalyst loading) were found to be of high relevance as confirmed by the respective F values (470.07 and 188.16). Moreover, the quadratic terms of MW (mixed quadratic, F = 250.12) and M^2 (pure quadratic, F =117.59) are also of equivalent significance as that of the linear terms. In general, the cubic terms were observed to be of marginally low significance, except MW^2 (F = 12.75).

Table 56.ANOVA for the esterification of FA and COFA

Source	Sum of squares	Degree of freedom(df)	Mean square	F value	p-Value for lack of fit
Model	5818.08	19	306.21	103.28	< 0.0001
M-M	1393.71	1	1393.71	470.07	< 0.0001
T-T	87.21	1	87.21	29.41	< 0.0001
W-W	557.87	1	557.87	188.16	< 0.0001
MT	2.65	1	2.65	0.89	0.3468
MW	741.58	1	741.58	250.12	< 0.0001
TW	12.95	1	12.95	4.37	0.0392
M^2	348.63	1	348.63	117.59	< 0.0001
T^2	140.31	1	140.31	47.32	< 0.0001
W^2	172.52	1	172.52	58.19	< 0.0001
MTW	7.37	1	7.37	2.49	0.1181
M^2T	8.68	1	8.68	2.93	0.0902
M^2W	1.49	1	1.49	0.50	0.4797
MT^2	8.59	1	8.59	2.90	0.0919
MW^2	37.79	1	37.79	12.75	0.0006
T^2W	3.04	1	3.04	1.03	0.3136
TW^2	4.37	1	4.37	1.47	0.2278
M^3	1163.84	1	1163.84	392.54	< 0.0001
T^3	40.01	1	40.01	13.50	0.0004
W^3	170.61	1	170.61	57.54	< 0.0001
Residual	290.56	98	2.96		
Cor Total	6108.64	117			

On the other hand, numerical proximity of R^2 (=0.95)and adjusted R^2 (=0.94)confirms the presence of non-redundant terms only in the model equation **[Equation 3]**. The final model equation is as follows:

$$\%conversion = +413.17069 + 296.3358M - 21.33893T - 8.22343W + 3.14973MT + 6.38627MW - 0.22210TW - 911.36269M^2 + 0.42433T^2 + 1.94962W^2 + 0.043668MTW - 0.70291M^2T - 0.71207M^2W - 0.023433MT^2 - 0.30078MW^2 - 1.36159 \times 10^{-3}T^2W + 4.07540 \times 10^{-3}TW^2 + 557.08005M^3 - 2.73927 \times 10^{-3}T^3 - 0.091741 \times W^3 \quad [3]$$

The 3D mesh plots, according to the cubic model are shown in **Figure 36 (a), (b) and (c).** Figure 36 (a) depicted the functional relation between % conversion (response), temperature and molar ration of FA: COFA. Both the trends were noted to be unimodal. Maximum % conversion was achieved at a molar ratio of 3.4:1 and at 60^0C reaction temperature. Upon increasing the molar ratio and the temperature beyond the optimum values, % conversion was observed to decrease again. This could be attributed to the increasing inhibition of the furfuryl alcohol and/or COFA over the stipulated parametric domain [140]. According to Kose et al 2002, similar result was obtained during the alcoholysis of cotton seed oil with methanol using the *Candida antarctica* as biocatalyst.

It may be noted that the molar ration was varied upon changing the amount of FA only. Therefore, the tentative explanation for the observed decrease in % conversion with varying amount of FA is the competitive inhibition by the reactants. It is often identified that alcohols, as well as fatty acids, tend to inhibit the lipase catalyzed reactions [136-139]. Hence, both the substrates act as dead-end inhibitors according to the experimental observations. Several authors [139-141] considered the ping-pong mechanism to describe

the enzymatic action of lipases, irrespective of the type of the biotransformation (i.e. esterification, interesterification, transesterification reactions).

$$FA + COFA \xleftrightarrow{K_{m(COFA)}} [E.COFA] \xrightarrow[H_2O]{} [E'] + FA \xleftrightarrow{K_{m(FA)}} [E'.FA] \rightarrow E + ester$$

$$K_{i(COFA)} \updownarrow +COFA \qquad\qquad K_{i(FA)} \updownarrow + FA$$

$$[E.COFA.COFA] \qquad\qquad [E'.FA.FA] \qquad [4]$$

The basic steps of this mechanism, as depicted above***[Equation 4]***, are assumed to be the following: COFA binds first to the enzyme forming a enzyme-COFA complex, which then isomerizes to an acyl-enzyme intermediate with the concomitant release of water. The second reactant, here alcohol reacts with acyl–enzyme complex to form another binary complex, which further isomerizes unimolecularly to a enzyme–ester complex that finally releases the ester and the enzyme. Both the alcohol and the acid, assumed to be competitive inhibitors, react with the lipase and the acyl–enzyme complex, respectively, to form dead-end complexes.

First, the alcohol acts on the acyl formed by nucleophilic attack on the fatty acid by the nucleophilic group at the active site of the lipase [140]. This alcohol reacts with the acylated enzyme leading to the ester formation and enzyme regeneration. Consequently, it is expected that the more nucleophilic an alcohol is, the faster its reaction rate. The nucleophily of alcohols is characterized by their electron pair donor numbers [143].According to a study, the effect of methanol enhanced by the known inductive effect of the methyl group that is more electron-releasing than hydrogen[143].However, it is assumed that this effect decreases with increasing distance to the reacting group. It becomes negligible after three to four σ-bonds, meaning that the alcohol nucleophilicity decreases with increasing number of carbon atoms (n), resulting in decreasing reactivities [increasing $K_{m(OH)}$ when n increases from 1 (methanol) to 5 (furfuryl alcohol)]. At the second level, where the alcohol acts on the enzyme reactivity, is the immediate reaction between the

alcohol and the enzyme, blocking nucleophilic site of the enzyme that, normally, is engaged in the acylation process. This might be responsible for the enzyme inhibition by the furfuryl alcohol. This process occurs between the electrophilic part of the alcohol that is the positively polarized hydrogen of the carboxylic group and the nucleophilic part of the enzyme.

Figure 36(a),(b) and (c) presented 3D Response surface plot of % conversion as function of (a) the temperature (45-65 ^{0}C) and molar ratio(1:1 to 5:1 of FA and COFA),(b)enzyme loading(4-12%w/w) and molar ratio(1:1 to 5:1 of FA and COFA) and (c) enzyme loading(4-12%w/w) and temperature(45-65 ^{0}C).

Figure 36 (a) depicted that the reaction temperature had relatively insignificant effect on % conversion compared to molar ratio. Above the stipulated optimum temperature (=60^{0}C), % conversion decreased as a result of reduced enzyme at higher temperatures. This result is in agreement with results obtained by Shimada et al. (1999), who conducted the methanolysis of vegetable oil using *Candida rugosa* lipase and methanol. It was reported that the ester conversion at 6 h increased with increasing temperature, with the optimal reaction temperature in the range of 50–60^{0}C.

Figure 36(b) described the effect of enzyme loading (% w/w) and molar ratio on % conversion. With the gradual increase in enzyme loading from 4 to 10 % w/w, % conversion was observed to increase at all values of molar ratio. This phenomenon confirmed the strong significance of enzyme loading on the response. Maximum % conversion was achieved at 10% enzyme loading and at a molar ratio of 3.4:1. It may be noted that temperature was held constant at its optimum value. As the trend was unimodal once again, % conversion above 10% w/w enzyme loading was recorded to decrease. Previous studies have reported similar observations, for example, in esterifications catalyzed by the *Candida antarctica* and *Rhizomucor meihei*[142-145].

Effect of variation in enzyme loading and temperatures on the % conversion is shown in the **Figure 36(c).** The trends were similar as described previously. Once again, the effect of temperature was found to be relatively insignificant compared to the enzyme loading. Similar studies have also demonstrated low influence of temperature on % conversion in the synthesis of oleic acid esters catalyzed by *R meihei* [80]. The optimum parametric condition, as depicted in Figure 36 is listed in **Table 57.**

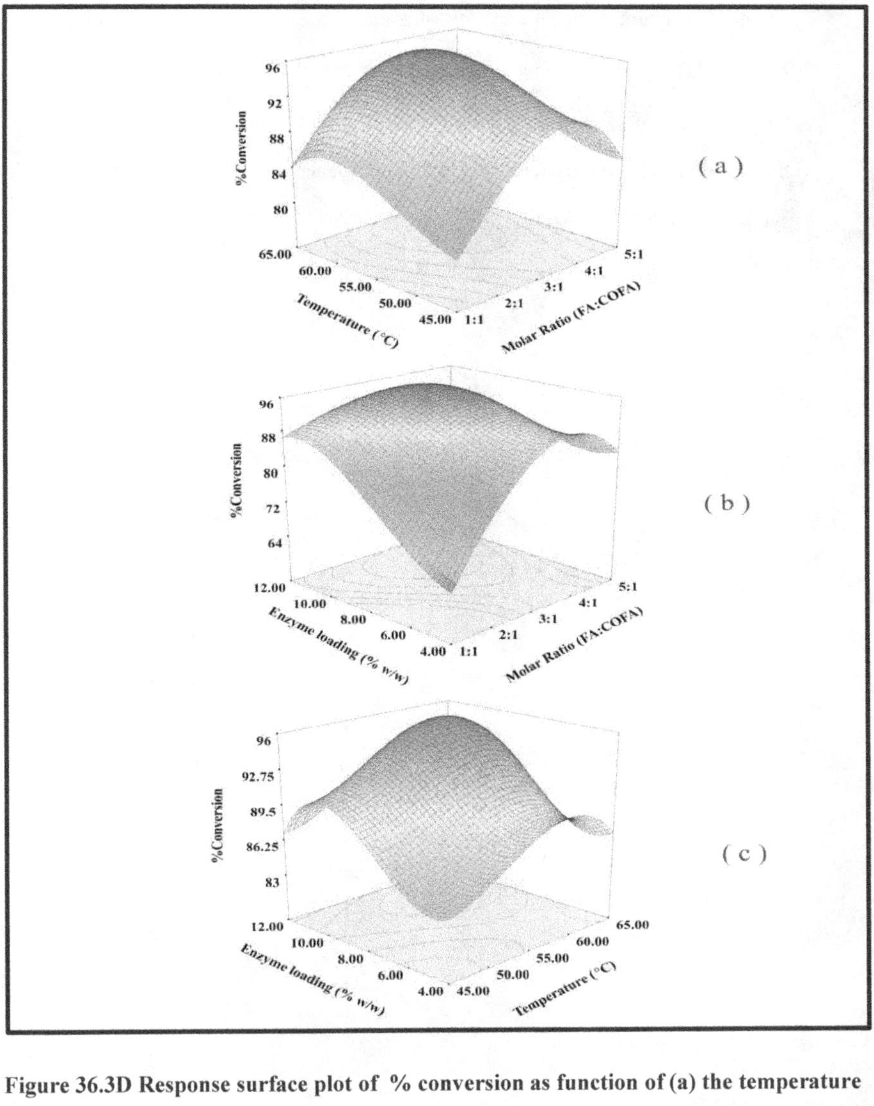

Figure 36.3D Response surface plot of % conversion as function of (a) the temperature (45-65 °C) and molar ratio(1:1 to 5:1 of FA and COFA),(b)enzyme loading(4-12%w/w) and molar ratio(1:1 to 5:1 of FA and COFA) and (c) enzyme loading(4-12%w/w) and temperature(45-65 °C).

Table 57.Optimum process parameters for the esterification

T (^{0}C)	W (%)	Molar Ratio	% Conversion	
			Actual	Predicted
60	10.21	3.4:1	96.00	99.00

In addition to the reaction equilibrium study, pertinent kinetic parameters according to the adopted ping pong bi-bi mechanism were also evaluated under optimum parametric condition by multivariable statistical regression.

The kinetic parameters include (i) the maximum rate of the reaction, V_0; the Michaelis constants for (ii) COFA, $K_{m(COFA)}$ and (iii) FA, $K_{m(FA)}$; the inhibition coefficients of (iv) COFA, $K_{i(COFA)}$ and (v) FA, $K_{i(FA)}$. The estimated parameter values are summarized in **Table 58**. High R^2 value (=0.96) clearly confirms the applicability of the adopted reaction scheme with inhibition. The diagram for cubic model has been provided in the **Figure 36**.

Table 58.Kinetic parameter for the esterification of furfuryl alcohol and castor oil fatty acid

Kinetic Parameters	Ping pong bi bi mechanism with inhibition
V_0(mol $l^{-1}h^{-1}$)	1.24
$K_{m(COFA)}$(mol l^{-1})	0.15
$K_{i(COFA)}$(mol l^{-1})	4.39
$K_{m(FA)}$ (mol l^{-1})	1.52
$K_{i(FA)}$ (mol l^{-1})	1.52
R^2	0.96

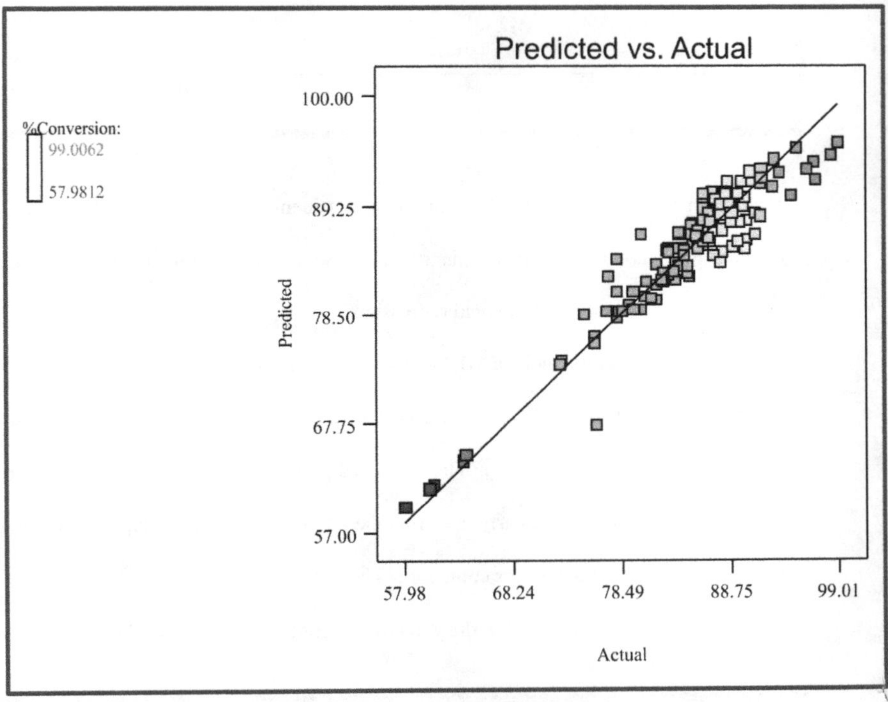

Figure 37.Actual versus predicted conversion (%) for Cubic Model

The observed inhibitory patterns in the present study appear to be same from the reported esterification kinetics, which indicate inhibition by FA and COFA. The inhibition reaction by short chain alcohol is well documented. In the present study, the inhibition observed by both FA and COFA is in concurrence with the following observations. However, the inhibition by substrate COFA can be explained by the hypothesis that the acyl-enzyme intermediate forms a dead end complex, thus the transfer of acyl moiety to alcohol is inhibited as shown in **equation 4** and the detailed inhibitory patterns were discussed earlier. Actual versus predicted % conversion for Cubic Model has been presented in the **Figure 37.**

Variation of the kinetic parameters with the changing reaction parameters (T, M and W)

The present study also investigates the variations of the reported kinetic parameters in the neighborhood of the optimal parametric condition set. The results are expected to be highly beneficial in constraint optimization of a continuous industrial reactor. It is worth noting that the optimal parametric conditions, evaluated from an experimental batch reactor, may not be implemented in toto in a continuous reactor. Precisely, this is because of system specific external constraints, such as (i) the total enzyme loading may undergo appreciable variation because of the corresponding demand-supply chain (ii) the energy budget may restrict the maximum allowable reactor temperature to be much below the optimum reaction temperature (iii) the supply of COFA may undergo severe variations because of any process upset in the upstream oil mill etc. Accordingly, the operating condition may get shifted from the optimal parametric condition. However, for the deviant operating point under external constraint, if the variations of the relevant kinetic parameters of the most appropriate rate equations are known a priori, one may decide about the local optima, which leads to maximum % conversion of the substrates under the external constraints. To our best of knowledge, variations of the kinetic parameters with the operating conditions have not been

reported till date. Profiles of the kinetic parameters, in the neighborhood of optimal condition are described in the **Figure 38 (a, b and c), Figure 39 (a, b and c), Figure 40 (a, b and c), Figure 41 (a, b and c) and Figure 42 (a, b and c)**. The characteristic features of the parameters are outlined as under:

According to **Figure 38 (a,b and c), Figure 39 (a,b and c), Figure 40 (a,b and c), Figure 41 (a,b and c) and Figure 42 (a,b and c)**, the variations of the kinetic parameters [V_0, $K_{m(COFA)}$, $K_{i(COFA)}$, $K_{m(FA)}$, and $K_{i(FA)}$] with the experimental parameters ,such as molar ratio, temperature and enzyme loading were investigated to interpret the respective trends of the reaction.

Variation in V_0

The contour plot for estimation of initial rate (V_0) of the esterification reaction of FA to COFA as a function of temperature, molar ratio and enzyme loading is depicted in **Figure 38(a),(b) and (c).** According to the initial trend, V_0 increased gradually with the l increase of all the process parameters (temperature, molar ratio and enzyme loading). However, V_0 passes through a maximum at a molar ratio of 3.94, temperature 45^0 C and an enzyme loading of 4 % w/w. This kind of phenomenon indicates that the enzyme activity is affected at higher FA content in the reaction mixture. Competitive inhibition might be the cause of decrease in the initial velocity of the reaction. It is often reported that alcohols, as well as fatty acids, tend to inhibit the lipase catalyzed reactions [136-137] as it was mentioned earlier.

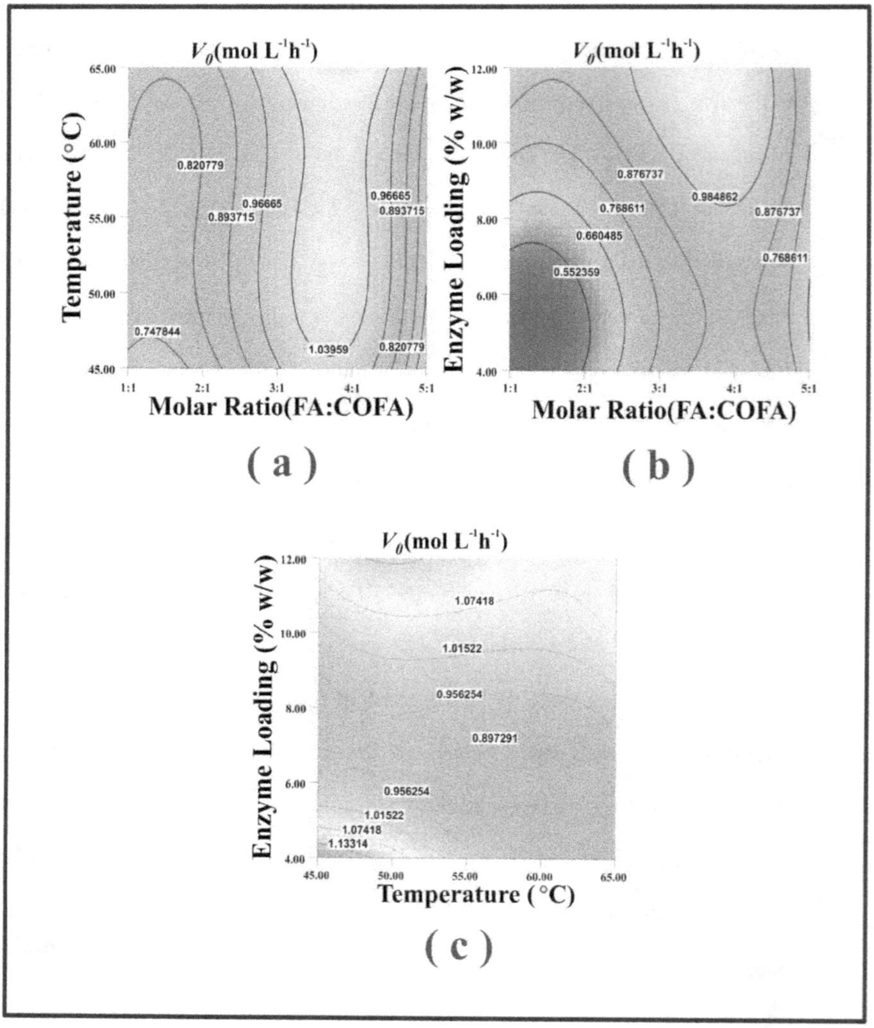

Figure 38. (a),(b) & (c) The contour plot for estimation of initial rate (V_0) of the esterification reaction of FA to COFA as a function of temperature, enzyme loading and molar ratio

Variation in $K_{m(COFA)}$

Figure 39 (a),(b) and (c) described the contour plots of Michaelis-Menten constant for COFA [$K_{m(COFA)}$]as a response of the independent regressor parameters. In accordance to the contour plot, $K_{m(COFA)}$ maximized at a molar ratio of 1,temperature 45°C and an enzyme loading of 12%w/w. Insignificant changes in the values of $K_{m(COFA)}$ were found to be observed with temperatures, though the trend was slightly decreasing. This phenomenon might be caused due to the enzyme inactivation at higher temperature. On the other hand, $K_{m(COFA)}$decreased with the increasing molar ratios . Competitive inhibition by the enzyme acyl complex might be the cause of decrease in the value of $K_{m(COFA)}$.

Increasing trend of the $K_{m(COFA)}$ was observed with increasing enzyme loading (%w/w),suggesting that the reaction might be kinetically controlled. The value of $K_{m(COFA)}$ gradually increased above 1:1 molar ratio of FA and COFA.As per the Michaelis -Menten hypothesis, the value of K_m increases with increasing substrate concentration. Similar kind of phenomenon was found to be observed in the previous studies of lipase catalyzed esterification reactions by *R.meihei* and *P. simplicissimum*[80,181].Hence, the lower temperature and higher enzyme loading can also be an option for the production of the ester under the external industrial constraints, when demand-supply chain of the enzyme will be hampered or the maximum allowable reactor temperature will be lower due to the energy budget restriction.

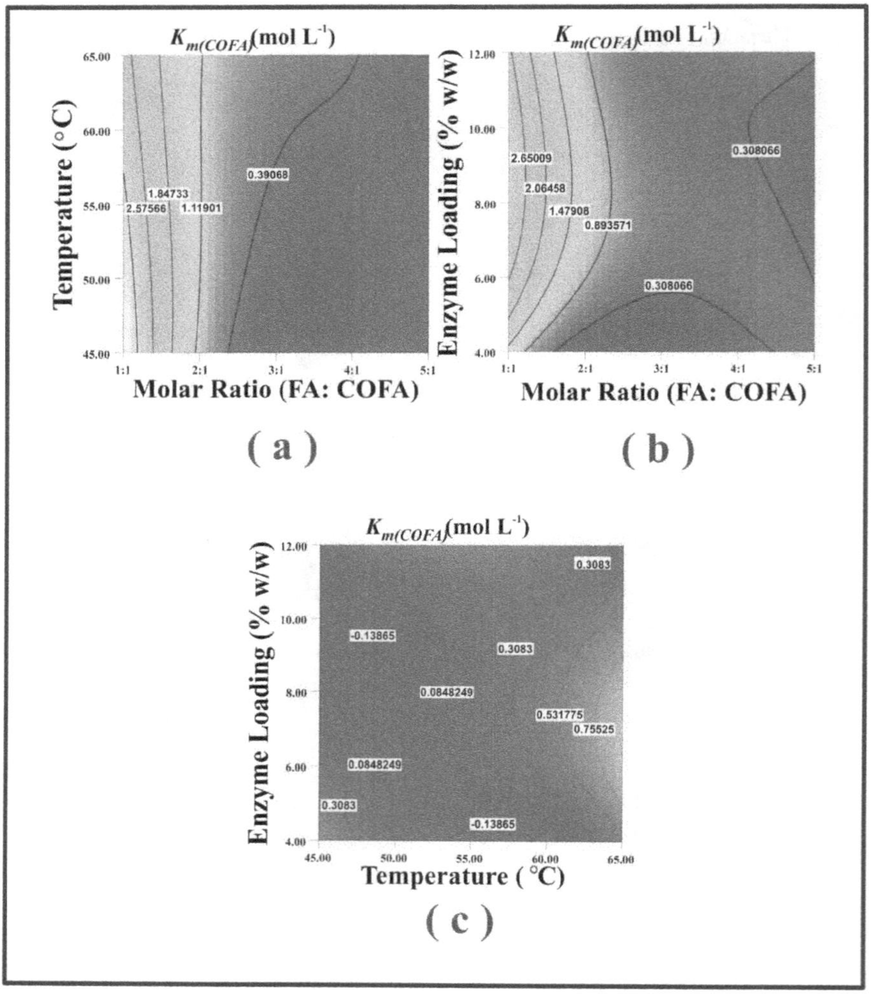

Figure 39 (a),(b) and (c) contour plots showing the effects of temperature, enzyme loading and molar ratio on Michaelis-Menten constant for COFA [$K_{m(COFA)}$]

Variation in $K_{i(COFA)}$

The combined effects of temperature, molar ratio and enzyme loading on the inhibition constant of COFA were found to be observed in the **Figure 40 (a),(b) and (c).**

The contour plots of $Ki_{(COFA)}$ revealed that, the inhibiton by COFA was found to be maximized at the molar ratio of 3.61 , at temperature of $45^{\circ}C$ and at an enzyme loading of 12%w/w. On the other hand the inhibition was found to be minimized at the molar ratio of 1.12,at $46.69^{\circ}C$ temperature and at an enzyme loading of 8.50%w/w. By comparing the *Km* and *Ki* values for the substrate COFA, some tendencies were observed. With the increasing values of $K_{m(COFA)}$ in the reaction system, the values of $K_{i(COFA)}$ were found to be decreased ,when the effects of various molar ratio and enzyme loading were evaluated. Here the Km and *Ki* are only dependent on the temperature (T).With the increasing temperature the value of $K_{i(COFA)}$ increased which means the reaction is slightly endothermic in nature.

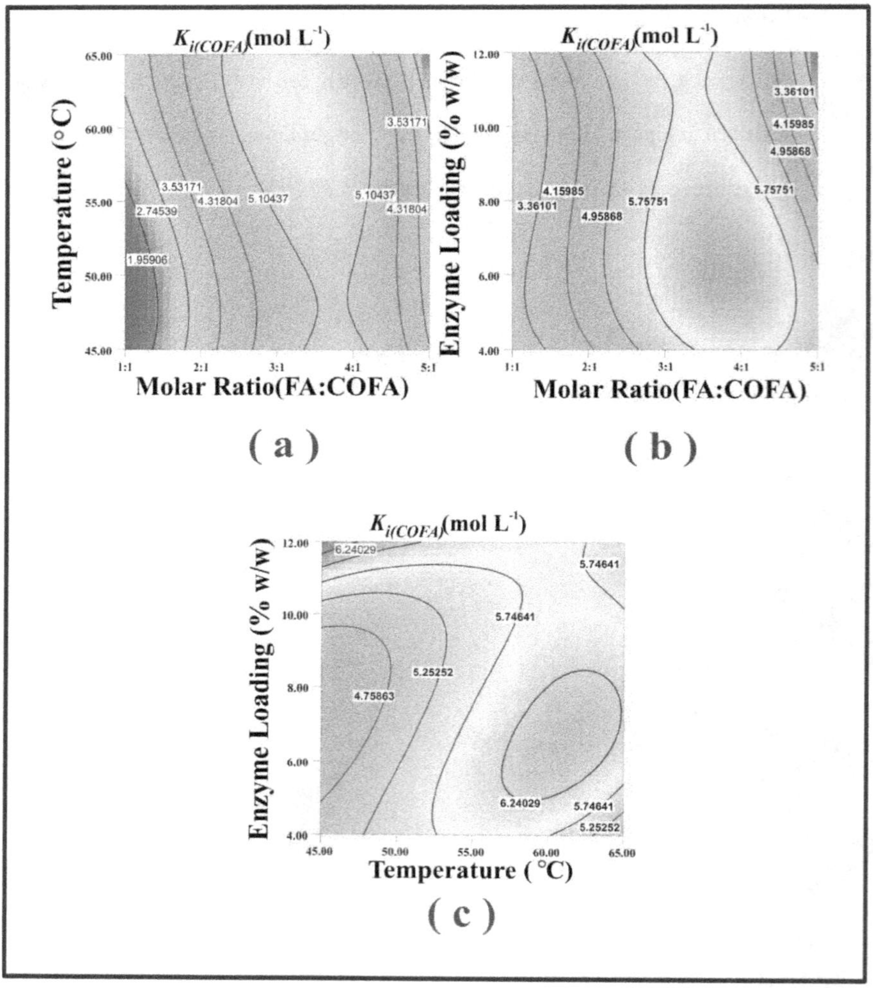

Figure 40 (a),(b) and (c) contour plots showing the effects of temperature, enzyme loading and molar ratio on inhibition constant for COFA [$K_{i(COFA)}$]

Variation in $K_{m(FA)}$

Furthermore the variations of Michaelis-Menten constant for the alcohol [$K_{m(FA)}$] are described in **Figure 41 (a), (b) and (c).** $K_{m(FA)}$ passes through a maximum at a molar ratio of 4.88,temperature 45.54°C and enzyme loading of 4% w/w .

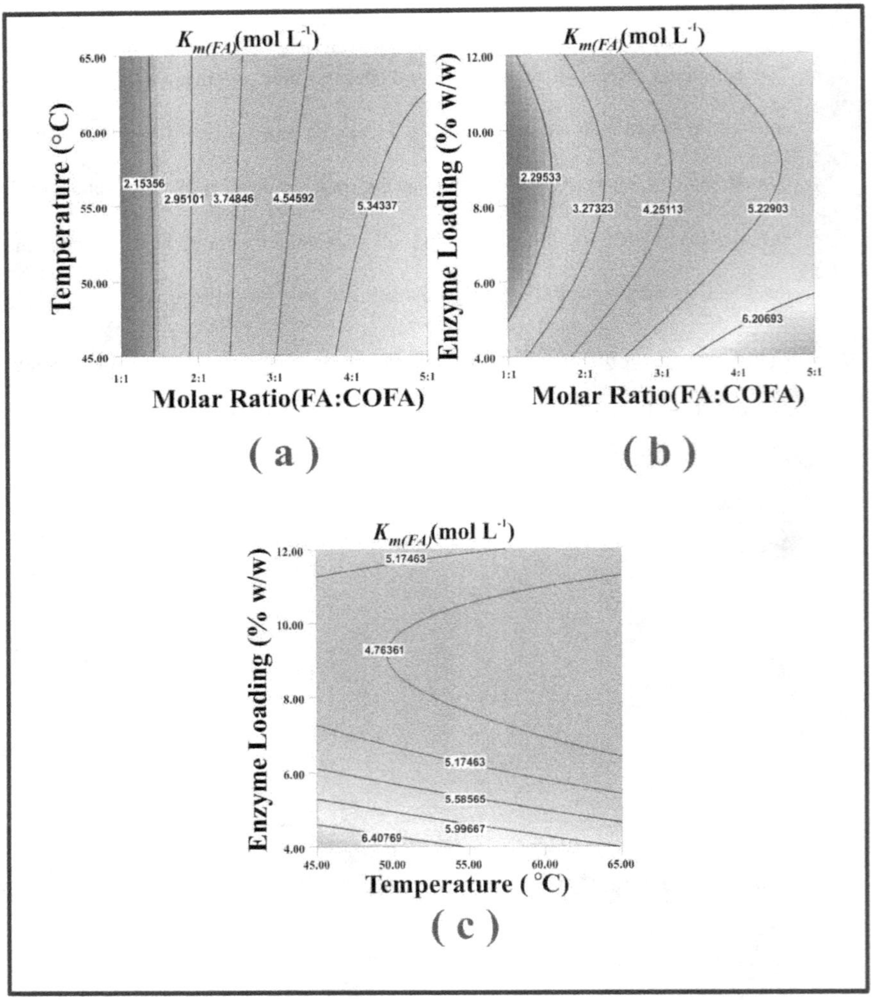

Figure 41 (a),(b) and (c) contour plots showing the effects of temperature, enzyme loading and molar ratio on Michaelis-Menten constant for the furfuryl alcohol [$K_{m(FA)}$]

Variation in $K_{i(FA)}$

The combined effects of temperature, molar ratio and enzyme loading on the inhibition constant of FA are shown in the **Figure 42 (a),(b) and (c)**.According to the trends of the contour plot, $K_{i(FA)}$ reaches at maxima at the molar ratio of 4.88,temperature 45.54°C and enzyme loading of 4%w/w. This kind of phenomenon might be happened due to the inactivation of enzyme at lower enzyme loading at lower temperature.

Hence the present investigation shows that the esterification kinetics in the reaction system follows a Ping Pong Bi Bi mechanism with competitive inhibition by FA and COFA.

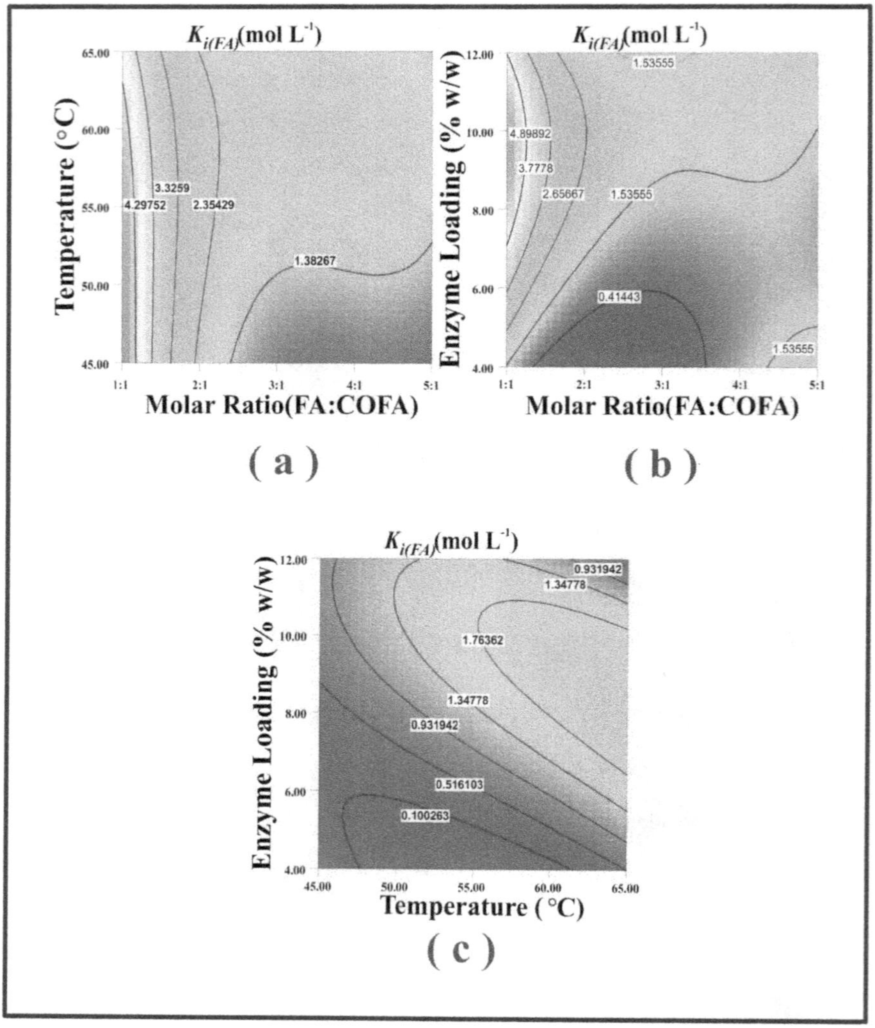

Figure 42 (a),(b) and (c) contour plots showing the effects of temperature, enzyme loading and molar ratio on inhibition constant FA [$K_{i(FA)}$]

CONCLUSION

In summary-

- The conversion of COFA into ester was significantly influenced by operating parameters; the most suitable reaction conditions for maximization of products were obtained at 3.4:1 molar ratio FA and COFA, at 60°C reaction temperature with 10.21%w/w enzyme loading.

- The reaction mechanism/kinetics for esterification reaction was confirmed by fitting the experimental data with rate equation given by Oliviera et. al and a good fit with R^2value of 0.96.

- Analysis of the kinetic data indicates that the esterification of FA with COFA catalyzed by free *Candida antarctica* lipase B can be described adequately by a Ping Pong bi–bi mechanism with competitive inhibition.

- Henceforth, the model and the kinetic constants obtained in this study can be used for innovative reactor design and process optimization under the constraint navigation of parametric space. Detail investigation of kinetic parameters, such as initial rate, equilibrium and inhibition coefficients etc. indicates the protocols of local optimization for esterification reaction of FA and COFA under restricted parametric condition (safety limits of temperature, low catalyst loading, non-optimal molar ratio etc), frequently encountered in industrial practice.

GENERAL CONCLUSION

The present study is concerned with the synthesis of plasticizer derived from vegetable oil and its refinery byproduct and its performance evaluation for the edible and industrial applications. The following conclusion emerged from the present research work:

1. In this study caprylic acid or capric acid were esterified separately with cetyl alcohol by enzymatic esterification reaction. Cetyl caprylate (CCAPL) and cetyl caprate (CCAPR) esters were successfully produced by the lipase catalyzed esterification reaction.

2. The CCAPL and CCAPR esters were successfully incorporated through emulsification process into the carboxymethyl cellulose (CMC) film blends as edible plasticizer in combination with conventional glycerol to improve barrier properties and mechanical properties. Mechanical properties were found to be increased with the increasing concentration of CCAPL and CCAPR ester into the CMC respectively. No significant effects were observed in case of crystalline nature and thermal properties of CCAPL and CCAPR ester containing CMC films.

3. Moisture barrier properties and water vapor permeability were also found to be improved with the addition of CCAPL and CCAPR ester individually in the CMC films. Therefore, it can be said that CCAPL and CCAPR esters consist the potential to be a good edible grade plasticizer for the edible coatings. Further CCAPL and CCAPR ester showed good antimicrobial activity against the bacterial strains such as *E. coli* and *S. aureus*. Henceforth, the application of CCAPL and CCAPR as edible plasticizer could also enhance the antimicrobial properties of any biopolymeric films.

4. Furfuryl alcohol (FA) and castor oil fatty acid (COFA) were successfully esterified via enzymatic esterification reaction. The ester yield was very significant at 60^0C temperature with 3:1 molar ratio of FA and COFA and 10% (w/w) immobilized lipase concentration for 5 h. The ester was found to be partially biodegradable to the environmental exposure up to a certain concentration.

5. The concerned FA-COFA ester obtained was successfully incorporated into PVC and EC films prepared by different solvent casting method, showing discrete migration into water. Mechanical tests in case of PVC film showed that the addition of FA-COFA ester as plasticizer resulted in a significant increase in elongation at break compared to pure PVC films, indicating the plasticization effect of this substance. Same phenomenon was found in case of ethyl cellulose film at the higher concentration of the FA-COFA ester. In both cases of PVC and EC films tensile properties were decreased with increase in FA-COFA ester content. No significant changes were observed in both the cases of unplasticized or plsticized PVC and EC films regarding thermal behavior and X-Ray diffraction. The incorporation of the FA-COFA ester in PVC and ethyl cellulose film showed promising results, indicating potential commercial application.

6. Furthermore the results of the tests in soil burial, enzymatic degradation as well as degradation in buffer solution showed the PVC polymer blends to be biodegradable. The biodegradability was found to be increased with the increase in the FA-COFA ester content. The FA-COFA ester content as high as 80% in the blend showed a high biodegradability of 47.58%.

7. Kinetic study includes the esterification reaction between FA and COFA using immobilized *Candida antartica* lipase B in a solvent free system. To study the occurrence of substrate inhibition, the effects of the lower molar ratio of FA and COFA on the initial reaction rate were investigated.

8. The conversion of COFA into ester was significantly influenced by variable reaction parameters; the most appropriate reaction conditions for maximization of products were obtained at 3.4:1 molar ratio FA and COFA, at 60°C reaction temperature with 10.21% (w/w) enzyme loading. Analysis of the kinetic data indicates that the esterification of FA with COFA catalyzed by *Candida antarctica* lipase B can be described adequately by a Ping Pong bi–bi mechanism with competitive inhibition.

CPSIA information can be obtained
at www.ICGtesting.com
Printed in the USA
LVHW042350100423
743983LV00010B/494

9 781805 252931